Lecture Notes in Control and Information Sciences

Edited by M. Thoma

Vol. 22: Optimization Techniques
Proceedings of the 9th IFIP Conference on
Optimization Techniques,
Warsaw, September 4–8, 1979
Part 1
Edited by K. Iracki, K. Malanowski, S. Walukiewicz
XVI, 569 pages. 1980

Vol. 23: Optimization Techniques
Proceedings of the 9th IFIP Conference on
Optimization Techniques,
Warsaw, September 4-8, 1979
Part 2
Edited by K. Iracki, K. Malanowski, S. Walukiewicz
XV, 621 pages. 1980

Vol. 24: Methods and Applications
in Adaptive Control
Proceedings of an International Symposium
Bochum, 1980
Edited by H. Unbehauen
VI, 309 pages. 1980

Vol. 25: Stochastic Differential Systems –
Filtering and Control
Proceedings of the IFIP-WG7/1 Working Conference
Vilnius, Lithuania, USSR, Aug. 28 – Sept. 2, 1978
Edited by B. Grigelionis
X, 362 pages. 1980

Vol. 26: D. L. Iglehart, G. S. Shedler
Regenerative Simulation of Response
Times in Networks of Queues
XII, 204 pages. 1980

Vol. 27: D. H. Jacobson, D. H. Martin, M. Pachter, T. Geveci
Extensions of Linear-Quadratic Control Theory
XI, 288 pages. 1980

Vol. 28: Analysis and Optimization of Systems
Proceedings of the Fourth International
Conference on Analysis and Optimization of Systems
Versailles, December 16–19, 1980
Edited by A. Bensoussan and J. L. Lions
XIV, 999 pages. 1980

Vol. 29: M. Vidyasagar,
Input-Output Analysis of Large-Scale
Interconnected Systems –
Decomposition, Well-Posedness and Stability
VI, 221 pages. 1981

Vol. 30: Optimization and Optimal Control
Proceedings of a Conference Held at
Oberwolfach, March 16–22, 1980
Edited by A. Auslender, W. Oettli, and J. Stoer
VIII, 254 pages. 1981

Vol. 31: Berc Rustem
Projection Methods in Constrained
Optimisation and Applications
to Optimal Policy Decisions
XV, 315 pages. 1981

Vol. 32: Tsuyoshi Matsuo,
Realization Theory of
Continuous-Time Dynamical Systems
VI, 329 pages, 1981

Vol. 33: Peter Dransfield
Hydraulic Control Systems –
Design and Analysis of Their Dynamics
VII, 227 pages, 1981

Vol. 34: H.W. Knobloch
Higher Order Necessary Conditions
in Optimal Control Theory
V, 173 pages, 1981

Vol. 35: Global Modelling
Proceedings of the IFIP-WG 7/1 Working
Conference Dubrovnik, Yugoslavia,
Sept. 1–5, 1980
Edited by S. Krčevinac
VIII, 232 pages, 1981

Vol. 36: Stochastic Differential Systems
Proceedings of the 3rd IFIP-WG 7/1
Working Conference
Visegrád, Hungary, Sept. 15–20, 1980
Edited by M. Arató, D. Vermes, A.V. Balakrishnan
VI, 238 pages, 1981

Vol. 37: Rüdiger Schmidt
Advances in Nonlinear
Parameter Optimization
VI, 159 pages, 1982

Vol. 38: System Modeling and Optimization
Proceedings of the 10 th IFIP Conference
New York City, USA, Aug. 31 – Sept. 4, 1981
Edited by R.F. Drenick and F. Kozin
XI, 894 pages. 1982

Vol. 39: Feedback Control of
Linear and Nonlinear Systems
Proceedings of the Joint Workshop
on Feedback and Synthesis of
Linear and Nonlinear Systems
Bielefeld/Rom
XIII, 284 pages. 1982

Vol. 40: Y.S. Hung, A.G.J. MacFarlane
Multivariable Feedback:
A Quasi-Classical Approach
X, 182 pages. 1982

Vol. 41: M. Gössel
Nonlinear Time-Discrete Systems –
A General Approach by
Nonlinear Superposition
VIII, 112 pages. 1982

Vol. 42: Advances in Filtering and
Optimal Stochastic Control
Proceedings of the IFIP-WG 7/1
Working Conference
Cocoyoc, Mexico, February 1–6, 1982
VIII, 391 pages. 1982

For information about Vols. 1–21 please contact your bookseller or Springer-Verlag.

Lecture Notes in Control and Information Sciences

Edited by M. Thoma

71

N. Baba

New Topics
in Learning Automata Theory
and Applications

Springer-Verlag
Berlin Heidelberg GmbH

Author
Norio Baba
Information Science and Systems Engineering
Faculty of Engineering
Tokushima University
Tokushima City, 770
Japan

ISBN 978-3-540-15613-0   ISBN 978-3-540-39497-6 (eBook)
DOI 10.1007/978-3-540-39497-6

Library of Congress Cataloging in Publication Data

Baba, N. (Norio).
New topics in learning automata theory and applications.
(Lecture notes in control and information sciences ; 71)
Bibliography: p.
Includes index.
1. Artificial intelligence.
I. Title.
II. Series.
Q335.B27   1984   001.53'5   85-10022

Originally published by Springer-Verlag Berlin Heidelberg New York in 1985.

2161/3020-543210

PREFACE

The appearance of a computer with huge memory is probably one of the most remarkable technological developments during the past two decades. We are now in the stage that sofisticated utilization of computers could make constructing an intelligent machine possible.

The study of artificial intelligence has been extensively done by many researchers. However, in spite of their efforts, its present state of development is still in its infancy. Active researches are now needed in order to utilize it for human welfare.

A concept of a learning automaton operating in an unknown random environment is one of the most important models that simulates an intelligent behavior of living beings. It was originally introduced by Tsetlin [T4], and since then, developed by many researchers. Since this model is fairly general, it would find various application areas.

This monograph presents some recent developments in the learning automata theory which are mainly concerned with the learning behaviors of stochastic automata under unknown multi-teacher environments. Although learning behaviors of stochastic automata have been considered quite extensively, almost all of the researches so far have dealt with only learning behaviors of stochastic automata under single teacher environment. Those researches should be extended in order to be applied to the problems (which we encounter considerably often) where one action elicits multi-responses from unknown multi-criteria environments. This monograph extends the researches having been obtained and deals with learning behaviors of stochastic automata under general multi-teacher environments.

Much of the research reported in this monograph is my recent work, and some part appears here for the first time. Chapter 2 deals with the learning behaviors of stochastic automata under unknown stationary multi-teacher environment. In Chapter 3, the learning behaviors of stochastic automata under nonstationary multi-teacher environment are discussed. Chapter 4 and Chapter 5 are concerned with the applications of the learning behaviors of stochastic automata. In particular, Chapter 4 deals with the parameter self-optimization problem with noise-corrupted, multi-objective functions as an application of learning behaviors of stochastic automata operating in an unknown nonstationary multi-teacher environment. Chapter 5 has no direct connections with the topics being dealt in this monograph. However, it deals with an application to the cooperative game by using the concept of the hierarchical structure automata which would become one of the most important tools in the near future. In the appendix, the learning behaviors of the hierarchical structure stochastic automata operating in the general multi-teacher environments are discussed.

If this monograph could make any contributions to the literature of learning automata and stimulate discussions among the researchers, it should give me a great pleasure.

It is a pleasure to acknowledge the encouragement of my teachers, Prof. Y. Sawaragi, Prof. T. Soeda, and Prof. T. Shoman. I am also indebted to my students Mr. H. Takeda and Mr. Y. Wajima for their assistance in preparing the manuscript. Finally, I would like to express my gratitude to my family, my father Yoshiyuki, my mother Fumiko, my wife Michiyo, and our children Hiroaki and Ayako for their encouragement and patience.

September 1984   Norio Baba,
                                                 University of Tokushima,
                                                 JAPAN.

CONTENTS

CHAPTER 1.   INTRODUCTION

1.1.   Introduction and Historical Remarks   1

1.2.   Outline of the Book   3

1.3.   Basic Model of the Stochastic Automaton Operating in
       a Single Teacher Environment   4

1.4.   Basic Norms of the Learning Behaviors of Variable-Structure
       Stochastic Automaton   7

1.5.   Several Representative Reinforcement Schemes and Their
       Learning Performances   9

1.6.   Appendix 1a - - - Some Background Material in Probability
       Theory   12

1.7.   Appendix 1b - - - Brief Comments about the Stochastic
       Processes Intrinsic to the Learning Behaviors of Stochastic
       Automata   15

CHAPTER 2.   LEARNING BEHAVIORS OF STOCHASTIC AUTOMATA UNDER MULTI-
             TEACHER ENVIRONMENT

2.1.   Introduction   17

2.2.   Basic Model   18

2.3.   Basic Norms of the Learning Behaviors of the Stochastic
       Automaton B in the General N-Teacher Environment   21

2.4.   Absolutely Expedient Nonlinear Reinforcement Schemes in
       the General N-Teacher Environment   25

2.5. Computer Simulation Results   37

2.6. Appendix 2a - - - Proof of the Lemma 2.6   50

2.7. Appendix 2b - - - Proof of the Lemma 2.7   52

CHAPTER 3.   LEARNING BEHAVIORS OF STOCHASTIC AUTOMATA UNDER

NONSTATIONARY MULTI-TEACHER ENVIRONMENT

3.1. Introduction   55

3.2. Learning Automaton Model under the Nonstationary Multi-
      Teacher Environment of S-model   56

3.3. ε-Optimal Reinforcement Scheme under the Nonstationary
      Multi-Teacher Environment   58

3.4. Computer Simulation Results   64

3.5. Comments and Concluding Remarks   69

CHAPTER 4.   APPLICATION TO NOISE-CORRUPTED, MULTI-OBJECTIVE PROBLEM

4.1. Introduction   71

4.2. Statement of the Problem   72

4.3. An Application of the Stochastic Automaton to the
      Noise-Corrupted, Multi-Objective Problem   72

4.4. Computer Simulation Results   78

4.5. Comments and Concluding Remarks   88

CHAPTER 5.   AN APPLICATION OF THE HIERARCHICAL STRUCTURE AUTOMATA

TO THE COOPERATIVE GAME WITH INCOMPLETE INFORMATION

5.1. Introduction   90

5.2. Statement of the Problem   91

5.3   Hierarchical Structure Stochastic Automata   91

5.4   An Application of the Hierarchical Structure Automata
       to the Cooperative Game   94

5.5   Computer Simulation Results   96

5.6   Comments and Concluding Remarks   103

5.7   Appendix - - - Learning Behaviors of the Hierarchical
       Structure Stochastic Automata Operating in the General
       Multi-Teacher Environments   104

Epilogue   109

References   110

Index   128

_ CHAPTER _1_

INTRODUCTION

1.1   Introduction and Historical Remarks

During the last three decades, the theory of optimal control has made
great progress.   It has reached a certain level of maturity.   However, in
order to apply this theory to actual problems, perfect information or a priori
information of the system must be known beforehand.   Therefore, the optimal
control theory sometimes cannot be applied to actual problems.   The idea of
learning control becomes necessary when information of the system is limited.
Recently, from such reason, the need of learning control has been accentuated.
Various approaches have been found useful for the objective of learning control.
In particular, learning automaton is one of the most important tools of learning
control.   Needless to say, if one wants to use learning automaton as the learning controller for an unknown system, one must investigate its learning performance in detail.

In this book, we will study the learning behaviors of stochastic automata
operating in the unknown general multi-teacher environment and consider applications to some practical problems.

Historically speaking, Tsetlin [T4] initially introduced the concept of
learning automaton operating in an unknown random environment.   He considered
learning behaviors of finite deterministic automata under the stationary random

environment $R(C_1, C_2, \ldots, C_r)$ and showed that they are asymptotically optimal under some conditions.

The study of learning behaviors of stochastic automata was started by Varshavskii and Vorontsova [V1] and since then have been studied quite extensively by many researchers. Norman [N15] considered stochastic automaton which has two states and showed that ε-optimality can be ensured by the L_{R-I} scheme.[*] Further, this scheme was proved to be ε-optimal in the general r-state case. [V8],[S3] Lakshmivarahan and Thathachar [L1] introduced the concept of absolutely expedient learning algorithms and proved that the algorithms in this class are also ε-optimal. Fu and Li [F6], Chandrasekaran and Shen [C1], and etc. also contributed fruitful results to the literature of learning automata. Survey papers written by Narendra and Thathachar [N3], Narendra and Lakshmivarahan [N4], and Narendra [N9] contain most of the recent work in this field along with the valuable comments for future research.

On the other hand, the applications of stochastic automata have also been considered by many researchers. McMurtry and Fu [M6] used stochastic automata for parameter self-optimization problem with unknown performance criteria. Shapiro and Narendra [S6], Viswanathan and Narendra [V7], Mason [M4], and Baba [B4] also considered this problem. Their studies suggest that the use of stochastic automata for this problem is quite efficient. The application of stochastic automata to the two person zero-sum games was tried by Chandrasekaran and Shen [C3]. Later, it was developed by Lakshmivarahan [L5]. Further, Waltz and Fu [W1] and Riordon [R2] used stochastic automaton as a learning controller of an

---

[*] See Appendix 2

unknown control system. Quite recently, the application of the learning automata theory to routing problems in communication networks was proposed by Mason [M3] and since then developed in great detail by Narendra et al [N5],[S10] Their attempts suggest a new course that researchers in this field should take in the future.

1.2   Outline of the Book

The problems discussed in this book are summarized as follows.

(1)   Learning behaviors of stochastic automata operating in the general unknown stationary multi-teacher environment

(2)   Learning behaviors of stochastic automata operating in a nonstationary multi-teacher environment

(3)   Some applications of stochastic automata

Chapter 2 is concerned with the learning behaviors of stochastic automata operating in the general unknown stationary multi-teacher environment. In order to discuss learning behaviors of stochastic automata under multi-teacher environment, the new concept of average weighted reward is introduced by considering a weighted average of the various responses from the multi-teacher environment. The definition of absolute expediency in the general n-teacher environment is given by using the newly introduced definition of the average weighted reward. The GAE scheme is proposed as a reinforcement algorithm of stochastic automaton. It is shown that this scheme ensures absolute expediency and ε-optimality in the general n-teacher environment.

Chapter 3 deals with the learning behaviors of stochastic automata operating in the nonstationary multi-teacher environment NMT from which stochastic automata receive responses having an arbitrary values between 0 and 1. As a generalized form of the GAE reinforcement scheme, the MGAE scheme is proposed. Further, it is shown that this scheme ensures ε-optimality in the nonstationary multi-teacher environment NMT.

Chapter 4 and Chapter 5 are devoted to the applications of stochastic automata. Chapter 4 is devoted to the parameter self-optimization problem with noise-corrupted, multi-objective functions by stochastic automata. It is shown that this problem can be reduced to that of the learning behaviors of stochastic automata operating in the nonstationary multi-teacher environment considered in Chapter 3. In Chapter 5, a coalition game between three players is considered. It is shown that the hierarchical structure stochastic automata are quite useful for finding an appropriate strategy in this game. In the appendix of this chapter, the learning behaviors of the hierarchical structure stochastic automata operating in the general multi-teacher environments are discussed.

1.3 Basic Model of the Stochastic Automaton Operating in a Single Teacher Environment

Figure 1 describes the learning mechanism of the stochastic automaton A operating in an unknown single teacher environment.*

The stochastic automaton A is defined by the sextuple $\{S,W,Y,g,P(t),T\}$. S denotes the set of two inputs $(0,1)$, where 0 indicates the reward response from $R(C_1,\ldots,C_r)$ and 1 indicates the penalty response. (If the set S consists of only two elements 0 and 1, the environment is said to be a P-model. When the

---

* The term "Unknown random environment" is synonymous with "Unknown teacher environment". (See Tsetlin [T4].)

input into A assumes a finite number of values in the closed interval $[0,1]$, it is said to be a Q-model. A S-model is one in which the input into A takes an arbitrary number in the closed line segment $[0,1]$.)

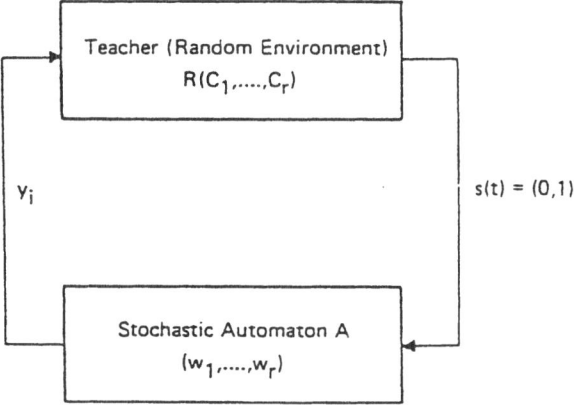

Figure 1 Basic model of a stochastic automaton

operating in an unknown teacher environment

W denotes the set of r internal states $(w_1,...,w_r)$.   Y denotes the set of r outputs $(y_1,...,y_r)$.   g denotes the output function $y(t) = g[w(t)]$, that is, one to one deterministic mapping.   P(t) denotes the probability vector $[p_1(t),...,p_r(t)]'$ at time t, and its ith component $p_i(t)$ indicates the probability with which the ith state w_i is chosen at time t. $(i=1,...,r)$

$$p_1(0) = ... = p_r(0) = 1/r, \quad \sum_{i=1}^{r} p_i(t) = 1$$

T denotes the reinforcement scheme which generates P(t+1) from P(t).

Suppose that the state w_i is chosen at time t.   Then, the stochastic automaton A performs action y_i on the random environment $R(C_1,...,C_r)$.   In response to the action y_i, the environment emits output s(t)=1 (penalty) with probability C_i and output s(t)=0 (reward) with probability $1-C_i$. $(i=1,...,r)$ If all the C_i $(i=1,...,r)$ are constant, the random environment $R(C_1,...,C_r)$ is said to be a stationary random environment (The term "single teacher environment" is also used instead of the term "a random environment".)   On the other hand, if C_i $(i=1,...,r)$ are not constant, it is said to be a nonstationary random environment.   Depending upon the action of the stochastic automaton A and the environmental response to it, the reinforcement scheme T changes the probability vector P(t) to P(t+1).

The values of C_i $(i=1,...,r)$ are not known a priori.   Therefore, it is necessary to reduce the average penalty,

$$M(t) = \sum_{i=1}^{r} p_i(t)C_i \tag{1}$$

by selecting an appropriate reinforcement scheme.

In the next section, several basic concepts useful for judging the effectiveness of variable-structure stochastic automata will be given.

1.4 Basic Norms of the Learning Behaviors of Variable-Structure

Stochastic Automaton

To judge the effectiveness of a learning automaton operating in a stationary single teacher environment $R(C_1, \ldots, C_r)$, various performance measures have been set up. In the following, let us briefly introduce these measures. We shall confine our discussions to the learning behaviors of stochastic automaton under a stationary single teacher environment of P-model. The learning behaviors of stochastic automaton under nonstationary multi-teacher environment of S-model will be discussed in Chapter 3.

Since the values of $C_i (i=1, \ldots, r)$ are not known a priori, the following definition can be derived.

Definition 1.1 A reinforcement scheme is said to be expedient if

$$\lim_{t \to \infty} E\{M(t)\} \quad < \quad \{ \frac{1}{r} \sum_{i=1}^{r} C_i \} \tag{2}$$

This definition means that a reinforcement scheme is said to be expedient if the average penalty in the limit is smaller than the initial average penalty. (Recall that $p_1(0) = p_2(0) = \ldots = p_r(0) = 1/r$.)

Assume that $C_\alpha = \min_i \{C_i\}$. Then, the optimal action of the stochastic automaton is y_α. Therefore, we can arrive at the following definitions.

Definition 1.2 A reinforcement scheme is said to be optimal if

$$\lim_{t \to \infty} E\{p_\alpha(t)\} \quad = \quad 1 \tag{3}$$

In spite of the great efforts of many authors, the general algorithm which ensures optimality has not been found so far. The following concept of ε-optimality has been introduced as a practical pinch hitter.

Definition 1.3 A reinforcement scheme is said to be ε-optimal if

$$\lim_{\theta \to 0} \lim_{t \to \infty} E\{p_\alpha(t)\} \quad = \quad 1 \tag{4}$$

where θ is a parameter included in the reinforcement scheme.

This definition implies that ε-optimality ensures the learning property of stochastic automaton which is very close to optimality. From definition 1.3, the following property can be derived.

For an arbitrary positive number ε, there exists some parameter θ_0 which ensures

$$\lim_{t \to \infty} E\{p_\alpha(t)\} \geq 1 - \varepsilon \qquad \text{for any } \theta \ (\ |\theta| < \theta_0 \ ).$$

Recently, Lakshmivarahan and Thathachar introduced the following concept of the absolute expediency.

<u>Definition 1.4</u> A reinforcement scheme is said to be absolutely expedient if

$$E\{M(t+1)/P(t)\} \quad < \quad M(t)$$

for all t, all $p_i(t) \in (0,1)$ $(i=1,\ldots,r)$, and all possible values of C_i $(i=1,\ldots,r)$. (The trivial case in which all the values of C_i $(i=1,\ldots,r)$ are equal is precluded.)

<u>Remark 1.1</u>   The definitions of optimality and ε-optimality can be easily transformed to the definitions described by $M(t)$. Let us give the definitions of optimality and ε-optimality by using $M(t)$.

<u>Definition 1.5</u> A reinforcement scheme is said to be optimal if

$$\lim_{t \to \infty} E\{M(t)\} \quad = \quad C_\alpha$$

<u>Definition 1.6</u> A reinforcement scheme is said to be ε-optimal if

$$\lim_{\theta \to 0} \lim_{t \to \infty} E\{M(t)\} \quad = \quad C_\alpha$$

where θ is a parameter included in the reinforcement scheme.

1.5 Several Representative Reinforcement Schemes and Their Learning

Performances

In the last section, we discussed several basic norms of the learning behaviors of stochastic automaton. In spite of the active efforts of many researchers, there have not been found so far any algorithm which ensures optimality in the general stationary random environment.

In this section, we will present several representative reinforcement schemes. The following L_{R-I} scheme is a well-known reward-inaction reinforcement scheme.

$\underline{L_{R-I}\ scheme}$

Assume that $y(t) = y_i$.

If $s(t) = 0$,

$$p_i(t+1)\ =\ (1 - \theta)p_i(t) + \theta, \qquad p_j(t+1)\ =\ (1 - \theta)p_j(t) \qquad (j \neq i)$$

If $s(t) = 1$,

$$p_m(t+1)\ =\ p_m(t) \qquad (\ m = 1,\ldots,r\ )$$

$$p_1(0) = \ldots = p_r(0) = 1/r, \qquad 0 < \theta < 1$$

The above reinforcement scheme has a nice learning property such as ε-optimality in the general stationary random environment $R(C_1,\ldots,C_r)$. This means that the L_{R-I} scheme ensures ε-optimality without any assumption about the values of C_i ($i=1,\ldots,r$).** However, the L_{R-I} scheme has also a drawback in the point that the state probability vector $P(t)$ is not altered when environmental response at time t is penalty $s(t) = 1$. (This means that the L_{R-I} scheme ignores penalty inputs from the random environment $R(C_1,\ldots,C_r)$.)

In 1973, Lakshmivarahan and Thathachar [L1] proposed the following general class of absolutely expedient learning algorithms which take penalty inputs from the random environment into account.

Absolutely Expedient Algorithm

Assume that $y(t) = y_i$.

If $s(t) = 0$,

$$p_i(t+1) = p_i(t) + \sum_{j \neq i} f_j(P(t)), \quad p_j(t+1) = p_j(t) - f_j(P(t)) \quad (j \neq i)$$

If $s(t) = 1$,

$$p_i(t+1) = p_i(t) - \sum_{j \neq i} g_j(P(t)), \quad p_j(t+1) = p_j(t) + g_j(P(t)) \quad (j \neq i)$$

**

In order to stress on the importance of this property, let us consider the following reinforcement scheme [S6].

Assume $y(t) = y_i$.

If $s(t) = 0$,

$$p_i(t+1) = p_i(t) + \theta p_i(t)(1 - p_i(t))$$

$$p_j(t+1) = p_j(t) - \theta p_i(t)(1 - p_i(t))/(r - 1) \qquad (j \neq i)$$

If $s(t) = 1$,

$$p_i(t+1) = p_i(t) - \theta p_i(t)(1 - p_i(t))$$

$$p_j(t+1) = p_j(t) + \theta p_i(t)(1 - p_i(t))/(r - 1) \qquad (j \neq i)$$

where $0 < \theta < 1$ and $p_1(0) = \ldots = p_r(0) = 1/r$.

It was shown in [S4] that the above algorithm ensures ε-optimality under the condition that one of the penalty probabilities is less than 1/2 and all others are greater than 1/2. Hence, if there is little a priori information about C_i, $i=1, \ldots, r$, we cannot expect that the above algorithm will attain ε-optimality.

They derived the following theorem :

Theorem 1.1   A necessary and sufficient condition for the stochastic automaton with the above reinforcement scheme to be absolutely expedient is

$$\frac{f_1(P(t))}{p_1(t)} = \ldots = \frac{f_r(P(t))}{p_r(t)} = \lambda(P) \tag{9}$$

$$\frac{g_1(P(t))}{p_1(t)} = \ldots = \frac{g_r(P(t))}{p_r(t)} = \mu(P) \tag{10}$$

where $\lambda(P)$ and $\mu(P)$ are arbitrary continuous functions satisfying

$0 < \lambda(P) < 1$ and $0 < \mu(P) < \min (p_j/(1-p_j))$ for all $j=1,\ldots,r$ and all p_j $(0,1)$.***

Remark 1.2   The L_{R-I} scheme is included in this class of algorithms.
(Let $f_j(P(t)) \triangleq \theta p_j(t)$ and $g_j(P(t)) \triangleq 0$.   Then, we can get the L_{R-I} scheme.)

Remark 1.3   As an example of the absolutely expedient algorithm, we can consider the following nonlinear reinforcement scheme.

Assume $y(t) = y_i$.

If $s(t) = 0$,

$p_i(t+1) = (1-\theta)p_i(t) + \theta$, $p_j(t+1) = (1-\theta)p_j(t)$ $(j \neq i)$

If $s(t) = 1$,

$p_i(t+1) = p_i(t) - k\theta(1-p_i(t))(H/(1-H))$, $p_j(t+1) = p_j(t) + k\theta p_j(t)(H/(1-H))$

$(j \neq i)$

$H = \min [p_1(t), \ldots , p_r(t)]$, $0 < \theta < 1$, $0 < k\theta < 1$

$p_1(0) = \ldots = p_r(0) = 1/r$.

***For notational convenience, we used the abbreviated forms $\lambda(P)$, $\mu(P)$, and p_j instead of $\lambda(P(t))$, $\mu(P(t))$, and $p_j(t)$ respectively.

1.6 Appendix 1a - - - Some Background Material in Probability Theory

The purpose of this appendix is to introduce some basic definitions, notations, and relations in the probability theory that are frequently used in this lecture notes. Although this appendix may be enough for understanding of this notes, it does not cover some of the fundamental materials of probability theory. Further, our introduction is by no means complete. (This is not a text on probability theory.) Readers desiring a more complete and detailed account of probability theory are refered to the books by Doob [D10], Loeve [L8], and etc.

Probability Measure Space

A probability measure space can be represented by the triple (Ω, B, P). Here, Ω denotes the basic space, B denotes the Borel field that has the following properties:

1) $\Omega \in B$ 2) If $A \in B$, then $\Omega - A \in B$

3) If $A_1, A_2, \ldots \in B$, then $\bigcup_{k=1}^{\infty} A_k \in B$

, and P denotes the probability measure that assigns a probability to each set in B. The function $P(\cdot)$ satisfies:

4) $P(A) \geq 0$ for every $A \ (\in B)$ 5) $P(\Omega) = 1$

6) $P(\bigcup_{k=1}^{\infty} A_k) = \sum_{k=1}^{\infty} P(A_k)$ for every mutually disjoint set $A_1, A_2,$

.... in B. $(A_i \cap A_j = \phi \ \ (i \neq j))$

Random Variables

A random variable can be defined as a measurable function on a probability measure space. This means that a function $x(\omega)$ ($x : \Omega \rightarrow R$), defined on the probability measure space (Ω, B, P), is called a (real) random variable if $\{\omega | \ x(\omega) \leq \lambda \} \in B$ for every real number λ.

Distribution Function and Probability Density Function

If $x(\omega)$ is a random variable defined on the probability measure space (Ω, B, P), then $\{ \omega \mid x(\omega) \leq \lambda \} \in B$ for every real number λ. Therefore, $P\{\omega \mid x(\omega) \leq \lambda\}$ can be defined for every real number λ.

Let $F(\lambda) \triangleq P\{\omega \mid x(\omega) \leq \lambda\}$.

The above function is called the distribution function of the random variable x.

It has the following properties:

(1) $F(\lambda)$ is the monotone non-decreasing and right-continuous function.

(2) $\lim\limits_{\lambda \to -\infty} F(\lambda) = 0$

(3) $\lim\limits_{\lambda \to +\infty} F(\lambda) = 1$

F has a probability density function f satisfying the following relation

$$F(\lambda) = \int_{-\infty}^{\lambda} f(\mu) d\mu \quad \text{if and only if F is absolutely continuous.}$$

Expectation

The mathematical expectation of the real random variable $x(\omega)$ is denoted by $E\{x(\omega)\}$ and is defined as follows.

$$E\{x(\omega)\} = \int_{\Omega} x(\omega) \, dP$$

This mathematical expectation can also be defined as follows by using the distribution function $F(\lambda)$ of the real random variable $x(\omega)$.

$$E\{x(\omega)\} = \int_{-\infty}^{+\infty} \lambda \, dF(\lambda)$$

Conditional Expectation

The conditional expectation is defined as follows.

Let x be a real random variable whose mathematical expectation exists and let be a Borel field. Let $\mathcal{F}'$ be the Borel field which includes all sets in $\mathcal{F}$ and any sets which differ from one of the sets in $\mathcal{F}$ by probability zero. The conditional expectation of x relative to $\mathcal{F}$ is denoted by $E\{x/\mathcal{F}\}$ and is defined

any measurable function (with respect to $\mathcal{F}'$) which satisfies the relation

$$\int_\Lambda E\{x/\mathcal{F}\} \, dP = \int_\Lambda x \, dP \qquad \text{for any set } \Lambda \text{ in } \mathcal{F}$$

Let $\Lambda \overset{\Delta}{=} \Omega$ in the above equality, then we can get

$$E\{E\{x/\mathcal{F}\}\} = E\{x\}$$

Stochastic Process

Let $(\Omega, \mathcal{F}, P)$ be the probability measure space. A stochastic process x

assigns to each time t in some set T a random variable $x_t(\omega)$ which is measurable

with respect to $\mathcal{F}$. If T is an infinite sequence, the stochastic process is

called a discrete parameter process. If T is an interval, $\{ x_t, t \in T \}$ becomes

a continuous parameter process. There are several important stochastic processes

to be noted, but we don't go into details. Readers having special interests to

stochastic processes can consult the books by Doob [D10], Dynkin [D11], and etc.

Of all the stochastic processes, Semi-Martingale may be one of the most

important stochastic processes in discussing the learning behaviors of stochastic

automata. This stochastic process can be defined as follows. (Doob [D10])

Let $\{ x_t, t \in T \}$ be a stochastic process with T, the set of infinite se-

quence, and $E\{|x_t|\} < \infty$, $t \in T$. Further, suppose that to each $t \in T$ corresponds

a Borel field $\mathcal{F}_t$ such that

1) $\mathcal{F}_s \subset \mathcal{F}_t \qquad s < t$

2) x_t is measurable with respect to $\mathcal{F}_t$ or is equal for almost all ω to

 a measurable function (with respect to $\mathcal{F}_t$).

The discrete parameter Semi-Martingale is the process which satisfies

$$x_s \leq E\{ x_t \mid \mathcal{F}_s \} \qquad (s < t) \qquad \text{with probability 1.}$$

The following convergence theorem is used quite frequently in this lecture notes.

For notational convenience, let T be the positive integer sequence beginning from 1. (T = 1,...)

Theorem (Doob [D10])   Let { x_t, $\mathfrak{F}_t$, t $\geq$ 1 } be a Semi-Martingale and let $\mathfrak{F}_\infty$ be the smallest Borel field including $\bigcup_{t=1}^{\infty} \mathfrak{F}_t$.   If the x_t's are uni- formly integrable, then $\lim_{t \to \infty} x_t = x_\infty$ exists with probability 1.

1.7   Appendix 1b - - - Brief Comments about the Stochastic Processes

Intrinsic to the Learning Behaviors of Stochastic

Automata

In the study of the learning behaviors of stochastic automata, our concern is often directed to the limiting behavior of $p_\alpha(t)$, a component of the state probability vector P(t) that corresponds to the least penalty probability C_α, or the average penalty $\sum_{i=1}^{r} p_i(t)C_i$.

In order to let readers be familiar with the stochastic process intrinsic to the learning behaviors of stochastic automata, let us consider the stochastic process induced by the learning behavior of the stochastic automaton with the L_{R-I} scheme under the nonstationary random environment $R(C_1(t,\omega),...,C_r(t,\omega))$ with the following property.

$$C_\alpha(t,\omega) + \delta < C_{k_1}(t,\omega), \quad . \quad . \quad . \quad , \quad C_{k_{r-1}}(t,\omega) \qquad \text{holds for some state}$$

w_α, some $\delta > 0$, all time t, and all ω ($\in \Omega$).

Here, the probability measure space (Ω, B, μ) can be defined as follows.

Let Ω be an arbitrary space.   Let B be the smallest Borel field including $\bigcup_{n=0}^{\infty} \mathfrak{F}_n$, where $\mathfrak{F}_n = \sigma(P(0),...,P(n),C(0),...,C(n))$.   ($\sigma(P(0),...,P(n),C(0),$ ...,C(n)) is the smallest Borel field of ω-sets with respect to which $p_1(0),$. ..,$p_r(0),p_1(1),...,p_r(1),....,p_1(n),...,p_r(n),C_1(0),...,C_r(0),....,C_1(n),...,$ and $C_r(n)$ are all measurable.)   It is clear from the definition that $\mathfrak{F}_n$ is the increasing Borel field with time n, that is to say,

$$\mathcal{F}_r \subset \mathcal{F}_t \qquad \text{if } r < t.$$

It can be easily shown that the stochastic process $\{p_\alpha(t)\}$ satisfies Semi-Martingale inequality:

$$E\{ p_\alpha(t+1) \ / \ \mathcal{F}_t \ \} \ \geq \ p_\alpha(t) \qquad \text{for all t.} \qquad (\text{ See Baba and Sawaragi}$$
[B1])

(For notational convenience, the conditional expectation

$E\{ p_\alpha(t+1) \ / \ \mathcal{F}_t \ \}$ is often replaced by $E\{ p_\alpha(t+1) \ / \ P(t) \ \}$.)

Since $p_\alpha(t)$ is uniformly bounded ($0 \leq p_\alpha(t) \leq 1$, it can be derived from Doob's theorem that $p_\alpha(t)$ converges with probability 1 to p_∞^α.

Remark A.1   From the above Semi-Martingale inequality, the following relation can be derived:

$$E\{p_\alpha(t+1)\} \ \geq \ E\{p_\alpha(t)\} \qquad \text{for all t.}$$

This means that the probability to select the optimal action y_α is monotonously increased in the sense of mathematical expectation.

Remark A.2   As having been discussed in the literature ([N15],[V8],[S3]), the L_{R-I} scheme has several nice learning properties in the general stationary random environment.   This might be the reason why the L_{R-I} scheme also behaves well in the above nonstationary random environment.   Therefore, a reinforcement scheme which cannot ensure any learning property in the stationary environment may not be able to behave well in the above nonstationary random environment.

LEARNING BEHAVIORS OF STOCHASTIC AUTOMATA

UNDER MULTI-TEACHER ENVIRONMENT

2.1   Introduction

In the last chapter, we have discussed learning behaviors of variable-
structure stochastic automata under a single teacher environment.   We have obtained several basic concepts such as "absolute expediency", "ε-optimality",
"optimality", and etc.   These concepts have played important roles in constructing efficient reinforcement schemes.   Together with advancement of the theory of
learning automata, various application areas have been found successfully.   The
application to routing of messages in communication network may be one of the
quite promissing examples.

However, the learning behaviors of stochastic automata under a single
teacher environment cannot be applied to the problems where one input elicits
multi-responses from the environment having multi-criteria.   When one considers
practical applications of learning automata, this problem is induced quite often.
Therefore, the research into the learning behaviors of stochastic automata under
unknown multi-teacher environment should be pushed forward.

Almost all research so far has dealt with learning behaviors of a single
automaton in a single teacher environment.   A quite few exceptional papers have

dealt with learning behaviors of stochastic automata operating in the multi-teacher environment. Recently, Koditschek and Narendra [K5] considered the learning behaviors of fixed-structure automata acting in a multi-teacher environment. Thathachar and Bhakthavathsalam [T1] then studied learning behaviors of variable-structure stochastic automata in two distinct teacher environment. The behavior of a collective of interacting stochastic automata in a single teacher environment was also considered by El-Fattah.

In this chapter, learning behaviors of variable-structure stochastic automata under general n-teacher environment will be considered.

2.2 Basic Model

The learning behaviors of stochastic automata in an unknown teacher environment have been discussed under the model shown in Figure 1. In this section, we generalize this model and discuss the learning behaviors of variable-structure stochastic automata operating in the general n-teacher environment, as illustrated in Figure 2.

Let us briefly explain the learning mechanism of the stochastic automaton B operating in the n-teacher environment.

The stochastic automaton B is defined by the set $\{S,Y,W,g,P(t),T\}$. S denotes the set of inputs $(i_1,i_2,\ldots,i_n)$ where i_j $(j=1,\ldots,n)$ is the response from the jth teacher $R_j(C_j^1,\ldots,C_j^r)$ and has binary values 0 and 1. 0 indicates the reward (nonpenalty) response from R_j, and 1 indicates the penalty response from R_j. Y denotes the set of r outputs $(y_1,\ldots,y_r)$. W denotes the set of r internal states $(w_1,\ldots,w_r)$. g denotes the output function $y(t) = g(w(t))$, that is, one to one deterministic mapping. $P(t)$ denotes the probability vector ($P(t) = ($ $p_1(t),\ldots,p_r(t))'$) at time t which governs the choice of the state. T denotes the reinforcement scheme which generates $P(t+1)$ from $P(t)$. The state w_k $(k=1,\ldots$

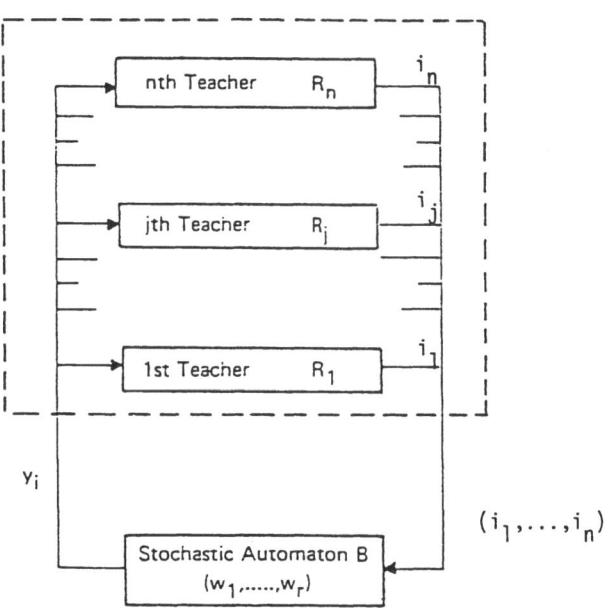

n-teacher environment

Figure 2   Stochastic automaton B operating in the

n-teacher environment

..,r) is chosen at time t with probability $p_k(t)$, where

$$p_1(0) = \ldots = p_r(0) = 1/r, \qquad \sum_{i=1}^{r} p_i(t) = 1.$$

Suppose that the kth state w_k is chosen at time t (k=1,...,r). Then, the stochastic automaton B performs action y_k on the n-teacher environment. In response to the action y_k, the jth teacher $R_j(c_j^1,\ldots,c_j^r)$ emits output s(t)=1 (penalty) with probability c_j^k and s(t)=0 (reward) with probability $1 - c_j^k$ (j=1,...,n). Depending upon the action of the stochastic automaton B and the n responses from the n-teacher environment, the reinforcement scheme T changes the probability vector from P(t) to P(t+1). The aim of the design is to choose reinforcement scheme T in such a manner that the stochastic automaton B improves its learning performance.

Remark 2.1   In the above model, the input set into stochastic automaton B is binary (0,1). In this case, the environment is said to be a P-model. If the input set into stochastic automaton B consists of a finite number of values in the closed interval [0,1], it is termed a Q-model. S-model is one in which the input into B can take an arbitrary value in the closed line segment [0,1]. In this chapter, we only deal with P-model. However, results which will be obtained can be easily extended to the Q- and S-model.

Remark 2.2   In this chapter, we consider the stationary multi-teacher environment in which the penalty probabilities c_j^k (k=1,...,r ; j=1,...,n) are all constant. Learning behaviors of stochastic automata under nonstationary multi-teacher environment of S-model will be discussed in the next section.

2.3 Basic Norms of the Learning Behaviors of the Stochastic Automaton

B in the General N-Teacher Environment

In chapter 1, various performance measures such as average penalty, opti-

mality, ε-optimality, and etc. have been set up. These measures must be modi-

fied when we consider the learning behaviors of stochastic automaton operating

in a multi-teacher environment.

In the following, let us give several basic norms of the learning behav-

iors of stochastic automaton operating in the general n-teacher environment

(GNTE) in which there exists a βth state w_β such that

$$c_1^\beta + \ldots + c_n^\beta \; < \; c_1^i + \ldots + c_n^i \qquad \text{for all} \; \; 1 \leq i \leq r \; (i \neq \beta) \qquad (1)$$

Remark 2.3   Koditschek and Narendra [K5] considered the fixed-structure

automata operating under the multi-teacher environment in which there is an

action y_γ such that

$$c_i^\gamma \; < \; c_i^j \qquad \text{for all i} \; (\; 1 \leq i \leq n \;) \; \text{and all j,} \qquad 1 \leq j \leq r, \; j \neq \gamma.$$
$$(2)$$

The condition (1) is more general than (2). Because it is assumed in (2) that

all the n-teachers agree that the γth action, y_γ is the best one.

As having been shown in chapter 1, the concept of the average penalty

has played quite important role to judge the effectiveness of a stochastic

automaton in a single teacher environment. In order to extend this concept to

the learning performance of stochastic automaton under multi-teacher environment

, the new definition should satisfy the following requirement. "When more

teachers agree with the action of stochastic automaton B, B receives greater

rewards from the n-teacher environment." From such a point of view, we arrive at the following definition of "the average weighted reward in the general n-teacher environment".

Definition 2.1   The average weighted reward in the n-teacher environment $W(t)$ is defined as follows:

$$W(t) = \sum_{i=1}^{r} [p_i(t)\{ \sum_{j=1}^{n} jD_{n,j}^{i}\}], \tag{3}$$

where $D_{n,j}^{i}$ means the probability that j teachers approve of the ith action y_i of the stochastic automaton B.   ($j = 1,\ldots,n$)   ($D_{n,j}^{i}$ can be described as follows:

$$D_{n,j}^{i} \triangleq \{(1-c_1^i)\ldots(1-c_j^i)\}\{c_{j+1}^i\ldots c_n^i\} + \ldots$$

$$\ldots + \{c_1^i\ldots c_{n-j}^i\}\{(1-c_{n-j+1}^i)\ldots(1-c_n^i)\}$$

$$(i = 1,\ldots,r \; ; \; j = 1,\ldots,n) \tag{4})$$

Remark 2.4   The average weighted reward $W(t)$ has the meaning of the average number of teachers out of n who would approve of the automaton action selected with the probability distribution $P(t)$.

Remark 2.5   Let $D_{n,0}^{i}$ be the probability that all n teachers disapprove of the ith action of the stochastic automaton B:

$$D_{n,0}^{i} \triangleq c_1^i c_2^i \ldots c_n^i, \quad i = 1,\ldots,r. \tag{5}$$

Then, clearly

$$\sum_{j=0}^{n} D_{n,j}^{i} = 1, \quad i = 1,\ldots,r. \tag{6}$$

Remark 2.6   When m teachers approve of the action y_i of stochastic automaton B, B gets m times as much reward as when only one teacher approves ($i=1,\ldots,r$).

Remark 2.7   If we extend the concept of average penalty straightly, we shall arrive at the notion of "average weighted penalty in the multi-teacher environment."   The above definition can be considered as a reverse concept.

Taking advantage of the above definition, we can introduce the new concept of "absolute expediency in the multi-teacher environment".

Definition 2.2   The stochastic automaton B is said to be "absolutely expedient in the general n-teacher environment" if

$$E\{W(t+1)/P(t)\} \; > \; W(t) \qquad \text{for all } t, \text{ all } p_i(t) \in (0,1), \; i=1,\ldots,r,$$
$$\text{and all } c_j^i \in (0,1), \; i=1,\ldots,r \; ; \; j=1,\ldots,n.$$

(7)

Remark 2.8   The above definition implies that the stochastic process $\{W(t), t=1,\ldots\}$ is a Semi-Martingale.   Taking mathematical expectations in both sides of the inequality (7), it can be easily shown that $E\{W(t)\}$ is monotonously increasing with time t. (See Appendix 1a.)   Therefore, the absolute expediency in the general n-teacher environment would be considered as a desirable and superior property.

Remark 2.9   The concept of the absolute expediency in a single teacher environment was initially suggested by Lakshmivarahan and Thathachar [L1].   The above definition can be considered as a generalized one.

Extending the notions of "expediency", "ε-optimality", "optimality" in the single teacher environment, we are able to define them in the general n-teacher environment.

Definition 2.3   The stochastic automaton B is said to be "expedient in the general n-teacher environment" if

$$\lim_{t\to\infty} E\{W(t)\} \; > \; W_0 \; , \text{ where } \; W_0 \; = \; \sum_{i=1}^{r} \frac{1}{r}\{ \sum_{j=1}^{n} jD_{n,j}^i \} \tag{8}$$

If no a priori information is available, the actions are chosen with equal probability and the value of the average weighted reward is equal to W_0. (W_0 is the initial average weighted reward.)   The above definition indicates that the average weighted reward is made larger than W_0, at least asymptotically.

Remark 2.10   It can be easily shown that absolute expediency in the general n-teacher environment implies expediency in the general n-teacher environment.

$$\text{Let} \quad W_\beta \triangleq \sum_{j=1}^{n} jD_{n,j}^{\beta} \tag{9}$$

The stochastic automaton B receives the maximal average weighted reward W_β when it chooses the action y_β with probability 1.   Therefore, we arrive at the following definition of optimality.

Definition 2.4   The stochastic automaton B is said to be "optimal in the general n-teacher environment" if

$$\lim_{t \to \infty} W(t) = W_\beta \quad \text{with probability 1.} \tag{10}$$

Optimality implies that asymptotically the action associated with the minimum sum of the penalty probabilities (See the condition (1).) is selected with probability one.   Therefore, this definition can also be given as follows.

Definition 2.4'   The stochastic automaton B is said to be "optimal in the general n-teacher environment" if

$$\lim_{t \to \infty} E\{p_\beta(t)\} = 1 \tag{11}$$

The following concept of ε-optimality can be used when the stochastic automaton B has the performance which is very close to optimality.

Definition 2.5   The stochastic automaton B is said to be "ε-optimal in the general n-teacher environment" if one can choose parameters included in the reinforcement scheme of the stochastic automaton B such that the following inequality (12) holds for any positive ε:

$$\lim_{t \to \infty} E\{W(t)\} \geq W_\beta - \varepsilon \tag{12}$$

This definition can also be represented as follows.

Definition 2.5   The stochastic automaton B is said to be "ε-optimal in the general n-teacher environment" if the parameters of the reinforcement scheme can be chosen so that

$$\lim_{t \to \infty} E\{p_\beta(t)\} \geq 1 - \epsilon \qquad \text{for any } \epsilon > 0. \tag{13}$$

2.4   Absolutely Expedient Nonlinear Reinforcement Schemes in the General N-Teacher Environment

Let us propose the following nonlinear reinforcement scheme (GAE scheme) in the general n-teacher environment.

GAE scheme :   When the output from the stochastic automaton B at time t is y_i and the responses from the multi-teacher environment are m rewards and (n-m) penalties, the state probability vector P(t) is transformed as follows :

$$P_i(t+1) = P_i(t) + (1 - \frac{m}{n})\{ \sum_{\substack{j \neq i \\ j=1}}^{r} \phi_j(P(t))\} - \frac{m}{n}\{ \sum_{\substack{j \neq i \\ j=1}}^{r} \psi_j(P(t))\} \tag{14}$$

$$P_j(t+1) = P_j(t) - (1 - \frac{m}{n})\phi_j(P(t)) + \frac{m}{n}\psi_j(P(t)) \qquad (1 \leq j \leq r; \ j \neq i) \tag{15}$$

where

$$\frac{\phi_1(P(t))}{P_1(t)} = \frac{\phi_2(P(t))}{P_2(t)} = \ldots = \frac{\phi_r(P(t))}{P_r(t)} = \lambda(P(t)) \tag{16}$$

$$\frac{\psi_1(P(t))}{P_1(t)} = \frac{\psi_2(P(t))}{P_2(t)} = \ldots = \frac{\psi_r(P(t))}{P_r(t)} = \mu(P(t)) \tag{17}$$

$$P_j(t) + \psi_j(P(t)) > 0, \qquad P_i(t) + \sum_{\substack{j \neq i \\ j=1}}^{r} \phi_j(P(t)) > 0,$$

$$p_j(t) - \phi_j(P(t)) < 1 \qquad (j = 1,\ldots,r \; ; \; i = 1,\ldots,r) \qquad (18)$$

Remark 2.11   The GAE reinforcement scheme is an extended form of the absolutely expedient learning algorithm [L1],[N4] in a single teacher environment. In [N4] let $x(n) \triangleq (n - m)/n$, then we can get (14) and (15).

Remark 2.12   The above algorithm describes a general form of the absolutely expedient reinforcement scheme in the general n-teacher environment as will be shown in the Theorem 1.   From the various examples of the GAE scheme, let us introduce the following two algorithms.

(1)   In (14) and (15) let $\phi_k \triangleq 0$ (k=1,...,r) and $\psi_k \triangleq - n\theta p_k$ (k=1,...,r), then we can get

$$p_i(t+1) \;=\; (1 - m\theta)p_i(t) + m\theta$$

$$p_j(t+1) \;=\; (1 - m\theta)p_j(t)$$

$$m = 0,\ldots,n; \;\; 1 \le j \le r; \;\; j \ne i; \;\; 0 < n\theta < 1.$$

Since the above algorithm is an extension of the reward-inaction scheme L_{R-I}, we shall call it GL_{R-I} scheme.

(2)   By letting $\psi_j(P(t)) \triangleq - \theta p_j(t)$ and $\phi_j(P(t)) \triangleq - k\theta(p_j(t))\{H/(1-H)\}$, (j=1,...,r), we can get the following reinforcement scheme.

$$p_i(t+1) \;=\; p_i(t) - k\theta(1 - \tfrac{m}{n})(1 - p_i(t))\{H/(1-H)\} + \theta(\tfrac{m}{n})(1 - p_i(t))$$

$$p_j(t+1) \;=\; p_j(t) + k\theta(1 - \tfrac{m}{n})(p_j(t))\{H/(1-H)\} - \theta(\tfrac{m}{n})(p_j(t)), \qquad (j \ne i)$$

$$0 < \theta < 1, \qquad H = \min[\, p_1(t),\ldots,p_r(t) \,], \qquad 0 < k\theta < 1.$$

We shall call this algorithm GNA scheme.

The learning behaviors of these two algorithms will be discussed in the later section.

Remark 2.13   The inequalities (18) certify that $p_k(t+1) \in (0,1)$ under the condition $p_k(t) \in (0,1)$ (k=1,...,r).

The GAE reinforcement scheme has several desirable learning performances. The following two theorems ensure absolute expediency and ε-optimality in the general n-teacher environment.

Theorem 2.1   Suppose that

$$\lambda(P(t)) \leq 0 \tag{19}$$

$$\mu(P(t)) \leq 0 \tag{20}$$

and

$$\lambda(P(t)) + \mu(P(t)) < 0 \tag{21}$$

for all t and P(t).

Then, the stochastic automaton with the GAE reinforcement scheme is absolutely expedient in the general n-teacher environment.

Theorem 2.2   Suppose that $\lambda(P(t)) = \theta(\overline{\lambda}(P(t))$ (22), $\mu(P(t)) = \theta\{\overline{\mu}(P(t))\}$ $(\theta > 0)$ (23), where $\overline{\lambda}(P(t))$ and $\overline{\mu}(P(t))$ are bounded functions which satisfy the conditions (19) $\sim$ (20) and $\overline{\lambda}(P(t)) + \overline{\mu}(P(t)) < - \widetilde{\varepsilon}$ for some positive number $\widetilde{\varepsilon}$. Then, the stochastic automaton B with the GAE reinforcement scheme is ε-optimal in the general n-teacher environment.

Since the proofs of the above two theorems are lengthy, we will begin by deriving several important lemmas.

Lemma 2.1:   Let

$$G_n^i \triangleq \sum_{j=1}^{n} jD_{n,j}^i \qquad (i=1,...,r) \tag{24}$$

, where $D_{n,j}^i$ is defined in (4).

Then, G_n^i can be simplified as follows.

$$G_n^i = n - (C_1^i + . . . + C_n^i) \qquad (i=1,...,r) \tag{25}$$

Proof: Let us use the mathematical induction method.

Let n = 1. Then clearly,

$$G_1^k = D_{1,1}^k = 1 - C_1^k. \qquad\qquad (k = 1,\ldots,r) \qquad\qquad (26)$$

Let n = 2. Then

$$G_2^k = 2D_{2,2}^k + 1D_{2,1}^k$$

$$= 2\{(1 - C_1^k)(1 - C_2^k)\} + \{(1 - C_1^k)C_2^k + C_1^k(1 - C_2^k)\}$$

$$= 2 - (C_1^k + C_2^k) \qquad\qquad (k = 1,\ldots,r) \qquad\qquad (27)$$

The above two equalities mean that (25) holds true for n = 1 and n = 2.

Now, let us assume that (25) holds true for n = n.

$$(\ G_n^k = n - (C_1^k + \ldots + C_n^k) \qquad\qquad (28) \qquad)$$

From (24),

$$G_{n+1}^k = (n+1)D_{n+1,n+1}^k + nD_{n+1,n}^k + \ldots + 1D_{n+1,1}^k. \qquad\qquad (29)$$

Clearly,

$$D_{n+1,i}^k = C_{n+1}^k D_{n,i}^k + (1 - C_{n+1}^k)D_{n,i-1}^k, \qquad 1 \leq i \leq n, \qquad\qquad (30)$$

$$D_{n+1,n+1}^k = (1 - C_{n+1}^k)D_{n,n}^k \qquad\qquad (31)$$

Therefore,

$$G_{n+1}^k = (n+1)(1 - C_{n+1}^k)D_{n,n}^k + \sum_{i=1}^{n} i\{C_{n+1}^k D_{n,i}^k + (1 - C_{n+1}^k)D_{n,i-1}^k\}$$

$$= (1 - C_{n+1}^k)\{nD_{n,n}^k + \sum_{i=1}^{n}(i-1)D_{n,i-1}^k\} + (1 - C_{n+1}^k)\{D_{n,n}^k + \sum_{i=1}^{n} D_{n,i-1}^k\}$$

$$+ C_{n+1}^k\{\sum_{i=1}^{n} iD_{n,i}^k\} \qquad\qquad (32)$$

It follows from (6), (24), and (28) that

$$G_{n+1}^k = (1 - C_{n+1}^k)G_n^k + (1 - C_{n+1}^k) + C_{n+1}^k G_n^k$$

$$= (G_n^k + 1) - C_{n+1}^k$$

$$= (n+1) - (C_1^k + \ldots + C_{n+1}^k) \tag{33}$$

This means that the equality (25) holds true for $n = n + 1$. Therefore, by means of mathematical induction method, we can show that the equality holds true for all n.

Q.E.D.

Lemma 2.2: The average weighted reward W(t) can be simplified as follows:

$$W(t) = n - \sum_{i=1}^{r} \{ (p_i(t))(C_1^i + \ldots + C_n^i) \}$$

Proof: From the definition of W(t) and the above lemma 2.1, we can easily get

$$W(t) = \sum_{i=1}^{r} \{ (p_i(t))G_n^i \}$$

$$= n - \sum_{i=1}^{r} \{ (p_i(t))(C_1^i + \ldots + C_n^i) \}$$

Q.E.D.

Lemma 2.3: The conditional expectation $E\{W(t+1)/P(t)\}$ can be described as follows:

$$E\{W(t+1)/P(t)\} = W(t) + \sum_{i=1}^{r} [(C_1^i + \ldots + C_n^i)(p_i(t) - E\{p_i(t+1)/P(t)\})]$$

Proof: This is obvious from the lemma 2.2.

Q.E.D.

Lemma 2.4: Suppose that $\phi_k(P(t))$ (k=1,...,r) and $\psi_j(P(t))$ (j=1,...,r) satisfy (16) and (17), respectively. Then,

$$p_i(t) - E\{p_i(t+1)/P(t)\} = \frac{p_i(t)\{\lambda(P(t)) + \mu(P(t))\}}{n}[(1-p_i(t))G_n^i - \sum_{\substack{j \neq i \\ j=1}}^{r} \{p_j(t)G_n^j\}],$$

$$(i = 1,\ldots,r) \tag{34}$$

<u>Proof:</u>   For the notational convenience we will often abbreviate time t.

$$\phi_i(P) \triangleq \phi_i(P(t)), \quad \psi_i(P) \triangleq \psi_i(P(t)), \quad p_i \triangleq p_i(t), \quad (\, i = 1,\ldots,r\,)$$

(35)

Further, let

$$\lambda \triangleq \lambda(P(t)), \qquad \mu \triangleq \mu(P(t)).$$

(36)

The conditional expectation $E\{p_i(t+1)/P(t)\}$ is calculated as follows.

$$E\{p_i(t+1)/P(t)\} = \sum_{m=0}^{n} p_i D_{n,m}^{i}[\, p_i + (1 - \frac{m}{n})\{ \sum_{\substack{j\neq i \\ j=1}}^{r} \phi_j(P)\} - \frac{m}{n}\{ \sum_{\substack{j\neq i \\ j=1}}^{r} \psi_j(P)\}\,]$$

$$+ \sum_{\substack{j\neq i \\ j=1}}^{r} p_j[\, \sum_{m=0}^{n} D_{n,m}^{j}[\, p_i - (1 - \frac{m}{n})\phi_i(P) + \frac{m}{n}\psi_i(P)]\,] \tag{37}$$

In (37), it is easily derived that

$$\sum_{m=0}^{n} p_i D_{n,m}^{i} p_i = p_i^{2} \tag{38}$$

$$\sum_{\substack{j\neq i \\ j=1}}^{r} p_j \sum_{m=0}^{n} D_{n,m}^{j} p_i = p_i(1 - p_i) = p_i - p_i^{2} \tag{39}$$

$$\sum_{m=0}^{n} p_i D_{n,m}^{i}(1 - \frac{m}{n}) \sum_{\substack{j\neq i \\ j=1}}^{r} \phi_j(P) = p_i\{ \sum_{\substack{j\neq i \\ j=1}}^{r} \phi_j(P)\}\{ \sum_{m=0}^{n} D_{n,m}^{i}(1 - \frac{m}{n})\}$$

$$= p_i\{ \sum_{\substack{j\neq i \\ j=1}}^{r} \phi_j(P)\}(1 - \frac{G_n^{i}}{n}) \tag{40}$$

$$\sum_{m=0}^{n} p_i D_{n,m}^{i}(- \frac{m}{n}) \sum_{\substack{j\neq i \\ j=1}}^{r} \psi_j(P) = - \frac{p_i}{n}\{ \sum_{\substack{j\neq i \\ j=1}}^{r} \psi_j(P)\} G_n^{i} \tag{41}$$

$$\sum_{\substack{j\neq i \\ j=1}}^{r} p_j [\sum_{m=0}^{n} D_{n,m}^j (-1 + \frac{m}{n})\phi_i(P)] = (- \phi_i(P))(1 - p_i) + \frac{\phi_i(P)}{n} \{ \sum_{\substack{j\neq i \\ j=1}}^{r} p_j G_n^j \} \quad (42)$$

$$\sum_{\substack{j\neq i \\ j=1}}^{r} p_j \sum_{m=0}^{n} D_{n,m}^j (\frac{m}{n}\psi_i(P)) = \sum_{\substack{j\neq i \\ j=1}}^{r} \frac{p_j}{n} \psi_i(P)G_n^j = \frac{\psi_i(P)}{n} \{ \sum_{\substack{j\neq i \\ j=1}}^{r} p_j G_n^j \} \quad (43)$$

Therefore, from (38) to (43), (37) can be transformed as follows.

$$E\{p_i(t+1)/P(t)\} = p_i + p_i\{ \sum_{\substack{j\neq i \\ j=1}}^{r} \phi_j(P)\}(1 - \frac{G_n^i}{n}) - \frac{p_i}{n}\{ \sum_{\substack{j\neq i \\ j=1}}^{r} \psi_j(P)\}G_n^i$$

$$- \phi_i(P)(1 - p_i) + \frac{\phi_i(P)}{n}\{ \sum_{\substack{j\neq i \\ j=1}}^{r} p_j G_n^j \} + \frac{\psi_i(P)}{n}\{ \sum_{\substack{j\neq i \\ j=1}}^{r} p_j G_n^j \} \quad (44)$$

From (16) and (17),

$$\phi_1(P) = \lambda p_1, \ldots , \phi_r(P) = \lambda p_r ;$$

$$\psi_1(P) = \mu p_1, \ldots , \psi_r(P) = \mu p_r. \quad (45)$$

From (44) and (45), we can get

$$E\{p_i(t+1)/P(t)\} = p_i - \frac{p_i(\lambda + \mu)}{n}[(1 - p_i)G_n^i - \{ \sum_{\substack{j\neq i \\ j=1}}^{r} p_j G_n^j \}] \quad (46)$$

Hence,

$$p_i(t) - E\{p_i(t+1)/P(t)\} = \frac{p_i(\lambda + \mu)}{n}[(1 - p_i)G_n^i - \{ \sum_{\substack{j\neq i \\ j=1}}^{r} p_j G_n^j \}]$$

Q.E.D.

Lemma 2.5   Suppose that the assumptions of Theorem 2.1 hold true. (See (19),
(20), and (21).)   Then, the GAE reinforcement scheme has the following learning
performance:

$$E\{p_\beta(t+1)/P(t)\} - p_\beta(t) \geq 0 \qquad \text{for all t and P(t)} \tag{47}$$

Proof: From Lemma 2.4,

$$E\{p_\beta(t+1)/P(t)\} - p_\beta(t) = \frac{- p_\beta(t)(\lambda + \mu)}{n}[(1 - p_\beta(t))G_n^\beta - \sum_{\substack{j \neq \beta \\ j=1}}^{r} p_j(t)G_n^j] \tag{48}$$

From Lemma 2.1,

$$G_n^i = n - (C_1^i + \ldots + C_n^i), \quad (i = 1,\ldots,r)$$

Therefore from (1),

$$G_n^\beta > G_n^j, \quad (j = 1,\ldots,r; j \neq \beta).$$

Consequently,

$$(1 - p_\beta(t))G_n^\beta - \sum_{\substack{j \neq \beta \\ j=1}}^{r} p_j(t)G_n^j \geq 0. \tag{49}$$

From the assumption of the lemma,

$$\lambda + \mu < 0. \tag{50}$$

Hence from (48), (49), and (50),

$$E\{p_\beta(t+1)/P(t)\} - p_\beta(t) \geq 0$$

(The equality holds when $p_\beta(t)$ = 0 or 1.)   Q.E.D.

Remark 2.14   Equation (47) is the Semi-Martingale inequality.   Taking the
mathematical expectations in both sides of (47), we can get

$$E\{p_\beta(t+1)\} \geq E\{p_\beta(t)\}, \text{ for all t} \tag{51}$$

This means that the mathematical expectation of $p_\beta(t)$ increases monotonously
with time t.

Using the lemmas which have been obtained, we can easily prove Theorem 2.1.

Proof of Theorem 2.1:   From Lemma 2.3 and Lemma 2.4,

$$E\{W(t+1)/P(t)\} = W(t) + [\sum_{i=1}^{r} \{(C_1^i+\ldots+C_n^i)(\frac{p_i(\lambda + \mu)}{n})\{(1 - p_i)G_n^i - \sum_{\substack{j \neq i \\ j=1}}^{r} p_j G_n^j\})\}] \tag{52}$$

Using Lemma 2.1, the above equality can be written as follows.

$$E\{W(t+1)/P(t)\} = W(t) + \frac{(\lambda + \mu)}{n}[\sum_{i=1}^{r} \{p_i(1-p_i)(-(C_1^i+\ldots+C_n^i)^2)$$

$$+ p_i(C_1^i+\ldots+C_n^i)(\sum_{\substack{j \neq i \\ j=1}}^{r} p_j(C_1^j+\ldots+C_n^j))\}] \tag{53}$$

Let

$$L \triangleq [\sum_{i=1}^{r} \{p_i(1-p_i)(-(C_1^i+\ldots+C_n^i)^2) + p_i(C_1^i+\ldots+C_n^i)(\sum_{\substack{j \neq i \\ j=1}}^{r} p_j(C_1^j+\ldots+C_n^j))\}]. \tag{54}$$

L can be simplified as

$$L = - \sum_{\substack{i \neq j \\ i < j \\ i,j}} \{p_i p_j((C_1^i+\ldots+C_n^i) - (C_1^j+\ldots+C_n^j))^2\} \tag{55}$$

Therefore from (1), (55), and the assumption of Theorem 2.1, we can get

$$E\{W(t+1)/P(t)\} > W(t)$$

for all t, all $p_i(t) \in (0,1)$, (i=1,...,r), and all $C_k^i \in (0,1)$,

i = 1,...,r, k = 1,...,n which satisfies the relation (1).

Q.E.D.

In order to prove Theorem 2.2, we need two more lemmas:

Lemma 2.6   Suppose that the assumptions of the Theorem 2.2 hold.   Then, $p_\beta(t)$ converges with probability 1.   Further, let $\lim_{t \to \infty} p_\beta(t) = p_\infty^\beta$ with probability 1. Then, $p_\infty^\beta = 1$ or 0 with probability 1.

Lemma 2.7   Suppose that the assumptions of the Theorem 2.2 hold.   Further, let

$$h_{x,\theta}(p) \overset{\Delta}{=} [\exp(xp/\theta) - 1]/[\exp(x/\theta) - 1], \quad (x > 0) \tag{56}$$

$$p_\beta'(t) \overset{\Delta}{=} 1 - p_\beta(t) \tag{57}$$

Then, there exists some positive number z which satisfies the inequality

$$E\{h_{z,\theta}(p_\beta'(t+1))/P(t)\} \leq h_{z,\theta}(p_\beta'(t)) \quad \text{for all t and P(t).} \tag{58}$$

Since the proofs of the above two lemmas are rather lengthy, we will find the space in the appendix.

Remark 2.15 $h_{z,\theta}(p)$ is a superregular function [N14],[N15] which satisfies the inequality $h_{z,\theta}(p) \geq Uh_{z,\theta}(p)$, where U is the operater $Uh_{z,\theta}(p(t)) = E\{h_{z,\theta}(p(t+1)/P(t)\}$.

Thanks to the above two lemmas(Lemma 2.6 and Lemma 2.7), Theorem 2.2 can be easily proved.

Proof of Theorem 2.2:

Let

$$\overline{p}_\infty^\beta \overset{\Delta}{=} 1 - p_\infty^\beta \tag{59}$$

Then from Lemma 2.6,

$$\overline{p}_\infty^\beta = 0 \quad \text{or} \quad 1 \quad \text{with probability 1.} \tag{60}$$

Furthermore,

$$0 < h_{z,\theta}(p) < 1 \qquad \text{when} \quad 0 < p < 1$$

$$h_{z,\theta}(0) = 0, \qquad h_{z,\theta}(1) = 1 \tag{61}$$

It follows from (60) and (61) that

$$\lim_{t\to\infty} h_{z,\theta}(p_\beta'(t)) = \overline{p_\infty^\beta} \qquad \text{with probability 1.} \tag{62}$$

Since $| h_{z,\theta}(p_\beta'(t)) |$ is bounded above (≤ 1),

$$\lim_{t\to\infty} \int_\Omega h_{z,\theta}(p_\beta'(t)) d\mu = \int_\Omega \lim_{t\to\infty} h_{z,\theta}(p_\beta'(t)) d\mu = \int_\Omega \overline{p_\infty^\beta} d\mu \tag{63}$$

From Lemma 2.7,

$$h_{z,\theta}(p_\beta'(0)) \geq \int_\Omega h_{z,\theta}(p_\beta'(1)) d\mu \geq \cdots$$

Consequently,

$$h_{z,\theta}(p_\beta'(0)) \geq \lim_{t\to\infty} \int_\Omega h_{z,\theta}(p_\beta'(t)) d\mu = \int_\Omega \overline{p_\infty^\beta} d\mu \tag{64}$$

It is shown in [N15] that

$$\lim_{\theta\to 0} h_{z,\theta}(p_\beta'(0)) = \lim_{\theta\to 0} \left[\frac{\exp(z(r-1)/r\theta) - 1}{\exp(z/\theta) - 1} \right] = 0 \tag{65}$$

Hence, from (64) and (65),

$$\lim_{\theta\to 0} E\{ \overline{p_\infty^\beta} \} = 0$$

This means that

$$\lim_{\theta\to 0} \lim_{t\to\infty} E\{ p_\beta(t) \} = 1 \qquad \text{Q.E.D.}$$

Absolute expediency and ε-optimality of the GAE reinforcement scheme have been proved under the very general condition (1) in which there exists one action y_β for which sum of the penalty probabilities is the least. This means that the learning of variable-structure stochastic automaton with the GAE scheme is done by a kind of majority decision.[*] For instance, let us consider the case in which there are some teachers who make mistakes, but almost all of the n-teachers agree on the best action y_δ. Even in such a case, the mathematical expectations of $p_\delta(t)$ and W(t) increase monotonously if sum of the penalty probabilities is the least for the action y_δ.[**] This means that the stochastic automaton with the GAE reinforcement scheme behaves well even in the multi-teacher environment in which some of n-teachers make mistakes.

As we mentioned earlier, Koditschek and Narendra [K5] considered learning behaviors of fixed-structure automata, and derived several fruitful results under the rather restrictive condition (2) in which all teachers agree that y_γ is the best action. Therefore, it seems that this property is the special character-istic of the variable-structure stochastic automata with the GAE reinforcement scheme.

---

[*] Strictly speaking, the learning of the variable-structure stochastic autom-aton with the GAE scheme is not done by a majority decision. We could consider the following example. "Even when only one teacher agrees on the action y_k and all of the other teachers have different opinions, the mathematical expectation of the probability $p_k(t)$ increases monotonously if sum of the penalty probabili-ties is the least for the action y_k."

[**] However, on the contrary, if there are many teachers who make mistakes and the sum of the penalty probabilities is not the least for the best action, then the stochastic automaton B will not be able to find the best action.

2.5   Computer Simulation Results

In the previous section, we showed that the GAE reinforcement scheme has nice learning properties such as absolute expediency and ϵ-optimality in the general n-teacher environment.   In this section, we present several computer simulation studies of the learning behaviors of the GL_{R-I} and GNA scheme (Remark 2.12) under multi-teacher environments.   Three examples are presented, the first one dealing with five state-stochastic automaton under three-teacher environment satisfying the condition (1), the second one dealing with five state stochastic automaton under five-teacher environment satisfying the condition (1), while the last one considers five-state stochastic automaton under five-teacher environment satisfying the more restrictive condition (2).   The first and second examples can be considered as examples in which some teachers agree on the best action but the other teachers make mistakes.   The last example can be considered as an example in which all teachers agree on the best action.

Example 1:   In this example, we consider the three-teacher environment in which two teachers (Teacher 1 and Teacher 3) agree on the best action ($\beta = 2$), but one teacher (Teacher 2) makes mistakes.   A learning behavior of the GL_{R-I} scheme under the three-teacher environment (TA3) with the characteristics described in Table I is simulated by computer and compared with those under the single-teacher environment (TA1) and the two-teacher environment (TA2). The changes in the probability $p_{\beta}(t)$ ($\beta = 2$) and the average weighted reward $W(t)$ are shown in Figure 3 and Figure 4, respectively.

Example 2:   In this example, we consider the five-teacher environment in which three teachers (Teacher 1, Teacher 2, and Teacher 4) agree on the best action (β = 1), but the other two teachers (Teacher 3 and Teacher 5) make mistakes.   A learning behavior of the GNA scheme under the five-teacher environment (TB5) with the characteristics described in Table II is simulated by computer and compared with those under the two single-teacher environments (TB1-1) and (TB1-3), the two-teacher environment (TB2), and the three-teacher environment (TB3).   The changes in the probability $p_\beta(t)$ (β = 1) and the average weighted reward W(t) are shown in Figure 5 and Figure 6, respectively.

Example 3:   In this example, we consider the five-teacher environment (TC5) that satisfies not only the condition (1), but also satisfies the more restrictive condition (2).   That is, all teachers agree that the third action is the best one.   (The penalty probabilities c_i^j (i = 1,...,5 ; j = 1,...,5) are described in Table III.)   A learning behavior of the GNA scheme under the five-teacher environment (TC5) is simulated by computer and compared with those under the single-teacher environment (TC1-1) and the three-teacher environment (TC3).   The changes in the probability $p_\beta(t)$ (β = 3) and the average weighted reward W(t) are shown in Figure 7 and Figure 8, respectively.

Table I

Three-Teacher Environment (TA3)

	c_i^1	c_i^2	c_i^3	c_i^4	c_i^5
Teacher 1	0.65	0.35	0.50	0.45	0.83
Teacher 2	0.35	0.38	0.40	0.85	0.95
Teacher 3	0.85	0.23	0.75	0.85	0.57
Sum of the penalty probabilities	1.85	0.96	1.65	2.15	2.35

(i = 1,2,3)

(TA1)	Single-teacher environment with Teacher 1
(TA2)	Two-teacher environment with Teacher 1 and Teacher 2
(TA3)	Three-teacher environment with Teacher 1 - Teacher 3

Table II

Five-Teacher Environment (TB5)

	c_i^1	c_i^2	c_i^3	c_i^4	c_i^5
Teacher 1	0.22	0.70	0.30	0.48	0.35
Teacher 2	0.17	0.57	0.68	0.55	0.47
Teacher 3	0.29	0.85	0.75	0.23	0.25
Teacher 4	0.18	0.83	0.62	0.79	0.72
Teacher 5	0.25	0.23	0.73	0.85	0.91
Sum of the penalty probabilities	1.11	3.18	3.08	2.90	2.70

$(i = 1,\ldots,5)$

(TB1-1)	Single-teacher environment with Teacher 1
(TB1-3)	Single-teacher environment with Teacher 3
(TB2)	Two-teacher environment with Teacher 1 and Teacher 2
(TB3)	Three-teacher environment with Teacher 1 - Teacher 3

Table III

Five-Teacher Environment (TC5)

	c_i^1	c_i^2	c_i^3	c_i^4	c_i^5
Teacher 1	0.72	0.62	0.18	0.55	0.66
Teacher 2	0.82	0.92	0.21	0.81	0.52
Teacher 3	0.45	0.55	0.25	0.72	0.82
Teacher 4	0.92	0.66	0.24	0.40	0.56
Teacher 5	0.61	0.83	0.22	0.74	0.82
Sum of the penalty probabilities	3.52	3.58	1.10	3.22	3.38

($i = 1,\ldots,5$)

(TC1-1)   Single-teacher environment with Teacher 1

(TC3)   Three-teacher environment with Teacher 1 - Teacher 3

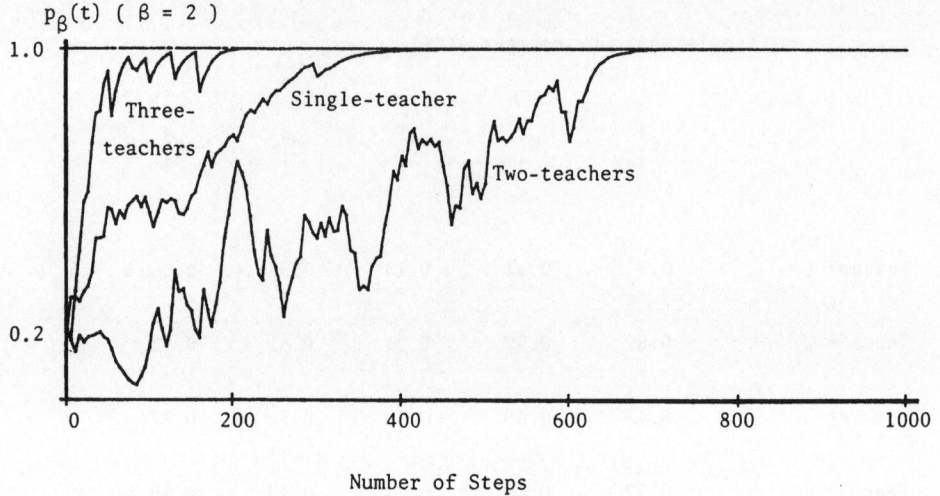

Figure 3   Changes in probability $p_\beta(t)$ in Example 1

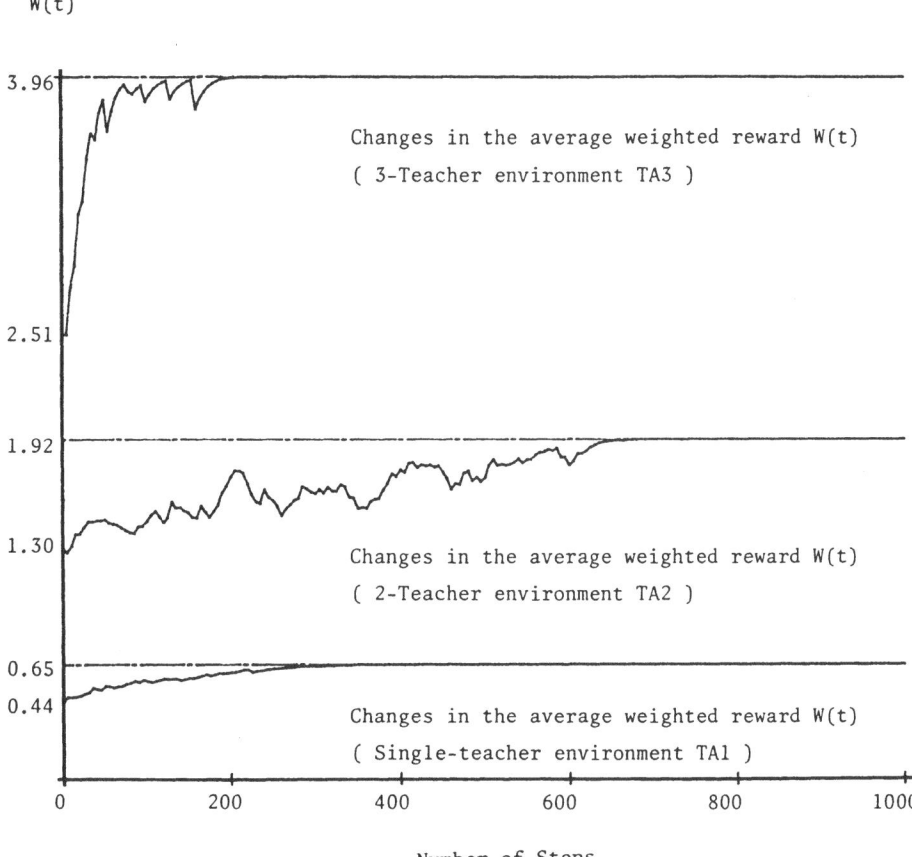

W(t)

3.96

Changes in the average weighted reward W(t)
(3-Teacher environment TA3)

2.51

1.92

1.30

Changes in the average weighted reward W(t)
(2-Teacher environment TA2)

0.65
0.44

Changes in the average weighted reward W(t)
(Single-teacher environment TA1)

0   200   400   600   800   1000

Number of Steps

Figure 4   Changes in the average weighted reward W(t) in Example 1

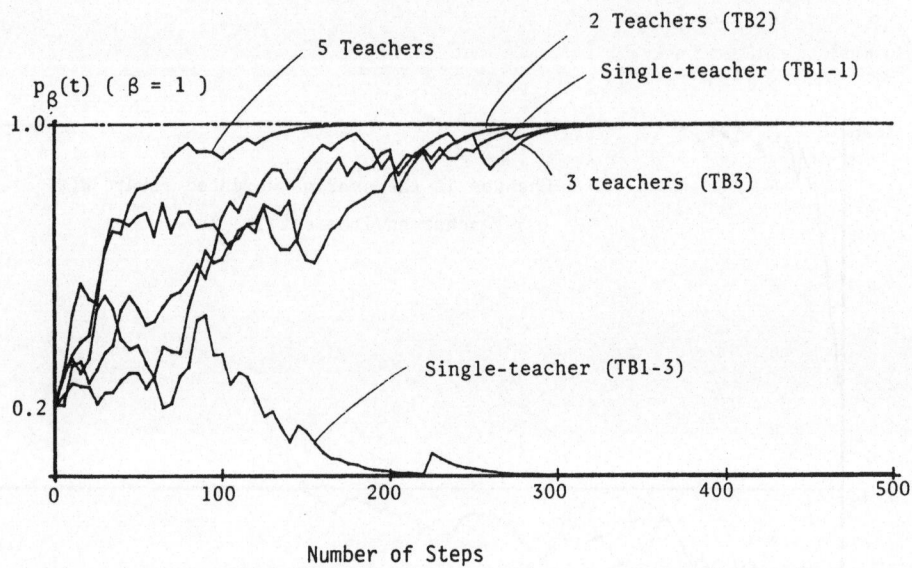

Figure 5   Changes in probability $p_\beta(t)$ in Example 2

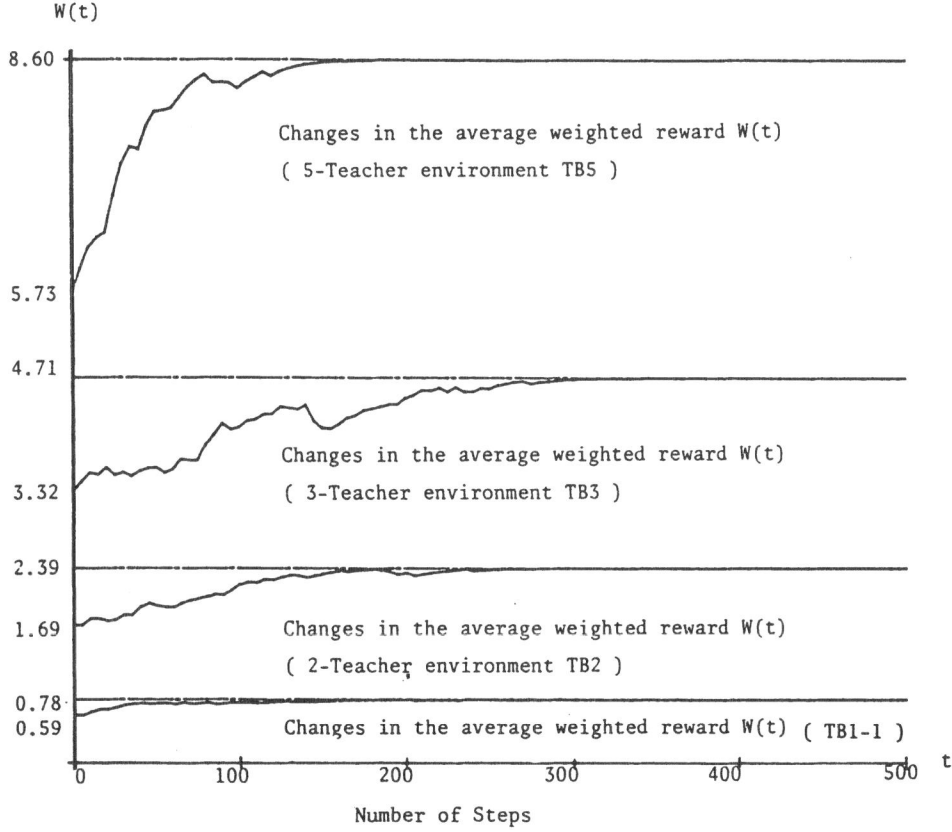

W(t)

8.60

Changes in the average weighted reward W(t)
(5-Teacher environment TB5)

5.73

4.71

Changes in the average weighted reward W(t)

3.32
(3-Teacher environment TB3)

2.39

1.69
Changes in the average weighted reward W(t)

(2-Teacher environment TB2)

0.78
0.59
Changes in the average weighted reward W(t) (TB1-1)

0   100   200   300   400   500   t

Number of Steps

Figure 6   Changes in the average weighted reward W(t) in Example 2

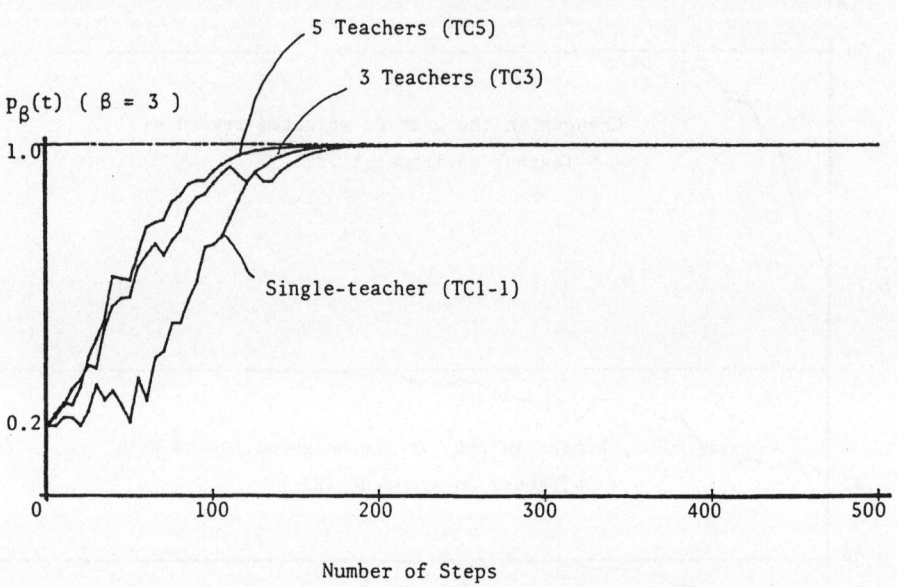

Figure 7   Changes in probability $p_\beta(t)$ in Example 3

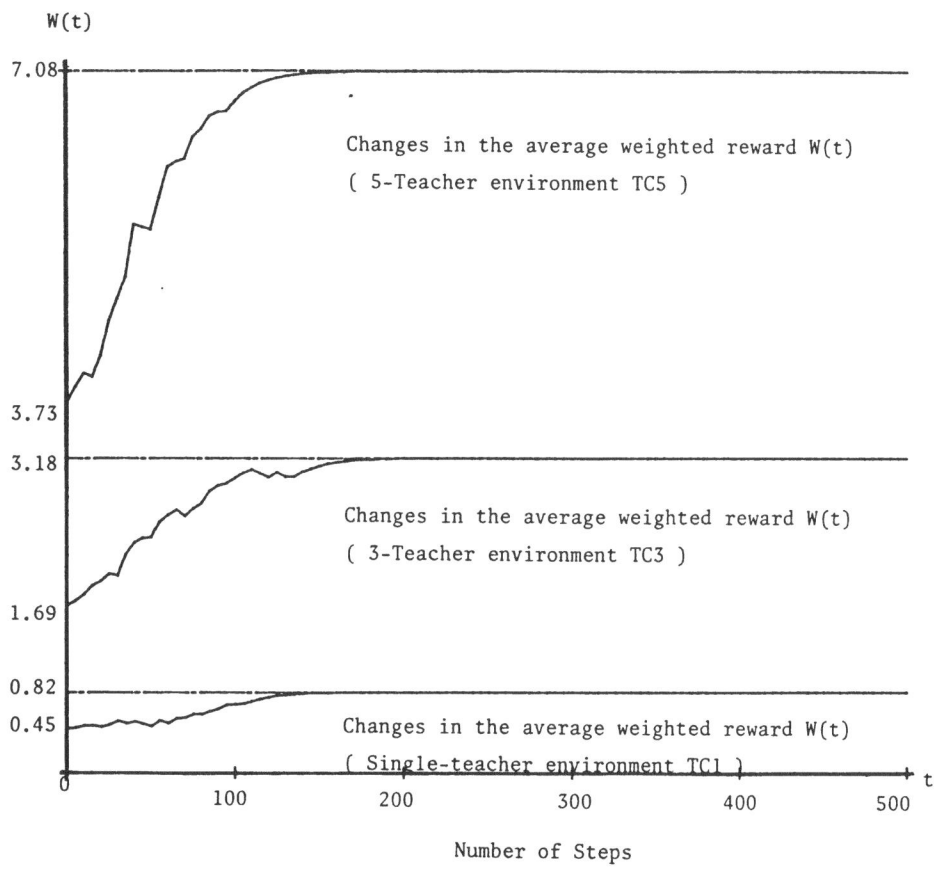

W(t)

7.08

Changes in the average weighted reward W(t)
(5-Teacher environment TC5)

3.73

3.18

Changes in the average weighted reward W(t)
(3-Teacher environment TC3)

1.69

0.82

0.45

Changes in the average weighted reward W(t)
(Single-teacher environment TC1)

t

100   200   300   400   500

Number of Steps

Figure 8   Changes in the average weighted reward W(t) in Example 3

Remark 2.16   In the first example, Teacher 2 makes a mistake.   Because Teacher 2 considers that y_1 is the best action. (In this example, y_2 is the best action.)   It appears that this is the reason why the speed of learning of the stochastic automaton under the two-teacher environment (TA2) is not faster than that under the single-teacher environment (TA1). (See Figure 3.) However, the speed of learning of the stochastic automaton under the three-teacher environment (TA3) is fastest since Teacher 3 agrees on the best action y_2.

Remark 2.17   In the second example, Teacher 3 and Teacher 5 make mistakes. Therefore, the stochastic automaton cannot find the best action y_1 under the single-teacher environment (TB1-3). (See Figure 5.)   Further, the speed of learning of the stochastic automaton under the three-teacher environment (TB3) is not necessarily faster than that under the two-teacher environment (TB2). It appears that this result comes from the fact that Teacher 3 makes a mistake but Teacher 1 and Teacher 2 agree on the best action y_1.

Remark 2.18   In the last example, all teachers agree on the best action y_3.   Therefore, the speed of learning of the stochastic automaton increases as more teachers are included.   (See Figure 7.)

Remark 2.19   As we mentioned earlier, the multi-teacher environment satisfying the condition (1) is more general than that satisfying (2).   We have considered three examples.   The first two examples satisfy only the general condition (1), but the last satisfies the more restrictive condition (2).   In the last example, all teachers have the same opinion that the third action is best.   It appears that this is the main reason why the speed of learning in Example 3 is faster than that in Example 2.

Remark 2.20   From the above three experimental results, we can derive

the following induction.   The GL_{R-I} and the GNA reinforcement scheme have a

nice convergence property which is very close to optimality.   We have obtained

the theoretical results in the previous section that the GAE reinforcement

scheme is absolutely expedient and ε-optimal in the general n-teacher environ-

ment.   The above computer simulation results confirm this theoretical study.

2.6   Appendix 2a — — —Proof of the Lemma 2.6

The first half of the Lemma 2.6 can be obtained from Lemma 2.5 and the fact that $p_\beta(t)$ is uniformly bounded ($0 \leq p_\beta(t) \leq 1$ for all t.).   It is also derived that

$$\lim_{t\to\infty} E\{p_\beta(t)\} \;=\; E\{p_\infty^\beta\} \qquad\qquad (\text{ Doob [D10] }) \tag{66}$$

Now, let us prove the latter half of this lemma.

Let

$$G_n^\delta \;\triangleq\; \max\; G_n^i \qquad (\; i = 1,\ldots,r \; ; \; i \neq \beta \;) \tag{67}$$

It is clear from (24) that

$$\sum_{j=1}^n C_j^\delta \;=\; \min \sum_{j=1}^n C_j^i \qquad (\; i = 1,\ldots,r \; ; \; i \neq \beta \;) \tag{68}$$

From (1), (25), and (68),

$$G_n^\beta \;>\; G_n^\delta \tag{69}$$

From the assumption of the Lemma 2.6 and (48),

$$E\{ \; p_\beta(t+1) \; / \; P(t) \; \} \; - \; p_\beta(t) \;\geq\; \frac{\theta \; \tilde{\epsilon} \; p_\beta(t)(1 - p_\beta(t))}{n}(\; G_n^\beta - G_n^\delta \;) \tag{70}$$

Assume now that there is a region S_1 such that $\mu(S_1) \neq 0$ and $0 < p_\infty^\beta < 1$ in S_1.

By taking the mathematical expectation in both sides of (70), we get

$$E\{p_\beta(t+1)\} - E\{p_\beta(t)\} \;\geq\; \int_\Omega \frac{\theta \; \tilde{\epsilon} \; p_\beta(1 - p_\beta)}{n}(\; G_n^\beta - G_n^\delta \;) \; d\mu \tag{71}$$

It can be concluded from (66) that

$$\lim_{t\to\infty} [\ E\{p_\beta(t+1)\} - E\{p_\beta(t)\}\ ]\ =\ \lim_{t\to\infty,} E\{p_\beta(t+1)\} - \lim_{t\to\infty} E\{p_\beta(t)\}\ =\ 0 \qquad (72)$$

Since $\dfrac{\theta\ \tilde{\varepsilon}\ p_\beta(1 - p_\beta)(G_n^\beta - G_n^\delta)}{n}$ converges with probability 1 and its absolute

value is bounded,

$$\lim_{t\to\infty} \int_\Omega \frac{\theta\ \tilde{\varepsilon}\ p_\beta(t)(1 - p_\beta(t))(G_n^\beta - G_n^\delta)}{n}\ d\mu$$

$$=\ \int_\Omega \frac{\theta\ \tilde{\varepsilon}\ p_\infty^\beta(1 - p_\infty^\beta)}{n}(G_n^\beta - G_n^\delta)\ d\mu$$

$$=\ \int_{S_1} \frac{\theta\ \tilde{\varepsilon}\ p_\infty^\beta(1 - p_\infty^\beta)}{n}(G_n^\beta - G_n^\delta)\ d\mu$$

$$>\ 0 \qquad\qquad\qquad\qquad\qquad\qquad\qquad\qquad\qquad\qquad (73)$$

It is clear from (71) that (72) is incompatible with (73).   Therefore,

$p_\infty^\beta\ =\ 1\ $ or $\ 0\ $ with probability 1.

Q.E.D.

2.7   Appendix 2b − − − Proof of the Lemma 2.7

The conditional expectation $E\{h_{x,\theta}(p'_\beta(t+1))/P(t)\}$ can be calculated as

$$E\{h_{x,\theta}(p'_\beta(t+1))/P(t)\}$$

$$= J\ [\ p_\beta(t)(\ \sum_{k=0}^{n} D^\beta_{n,k}\{\exp[x\{1 - (\ p_\beta(t) + (1 - \frac{k}{n})\sum_{\substack{j\neq\beta \\ j=1}}^{r}\phi_j - \frac{k}{n}\sum_{\substack{j\neq\beta \\ j=1}}^{r}\psi_j)\}/\theta] - 1\})$$

$$+ \sum_{\substack{i\neq\beta \\ i=1}}^{r} p_i(t)(\ \sum_{k=0}^{n} D^i_{n,k}\{\exp[x\{1 - (p_\beta(t) - (1 - \frac{k}{n})\phi_\beta + \frac{k}{n}\psi_\beta)\}/\theta] - 1\})\ ] \quad (74)$$

, where $J \overset{\Delta}{=} 1/[\exp(x/\theta) - 1]$, $p_\beta \overset{\Delta}{=} p_\beta(t)$, and $D^i_{n,k}$ ($i = 1,\ldots,r$; $k = 0,\ldots,n$) are defined in (4).

Taking into account the relation (6), we can get

$$E\{h_{x,\theta}(p'_\beta(t+1))/P(t)\}\ =\ J\ [\ \{\exp(xp'_\beta(t)/\theta)\}\{M(x,P)\} - 1\ ] \quad (75)$$

, where

$$M(x,P) = p_\beta(t)(\ \sum_{k=0}^{n} D^\beta_{n,k}\{\exp[\{- x(1-\frac{k}{n})\sum_{\substack{j\neq\beta \\ j=1}}^{r}\phi_j + \frac{k}{n} x \sum_{\substack{j\neq\beta \\ j=1}}^{r}\psi_j\}/\theta]\})$$

$$+ \sum_{\substack{i\neq\beta \\ i=1}}^{r} p_i(t)(\ \sum_{k=0}^{n} D^i_{n,k}\{\exp[\{x(1 - \frac{k}{n})\phi_\beta - \frac{k}{n} x\ \psi_\beta\}/\theta]\}) \quad (76)$$

From (16), (17), (22), and (23),

$$M(x,P) = p_\beta(t)(\ \sum_{k=0}^{n} D^\beta_{n,k}\{\exp[\ - x(1 - \frac{k}{n})\overline{\lambda}(1 - p_\beta) + \frac{k}{n} x\ \overline{\mu}(1 - p_\beta)\ ]\}\ )$$

$$+ \sum_{\substack{i\neq\beta \\ i=1}}^{r} p_i(t)(\ \sum_{k=0}^{n} D^i_{n,k}\{\exp[x(1 - \frac{k}{n})\overline{\lambda}p_\beta - \frac{k}{n} x\ \overline{\mu}\ p_\beta]\}\ ) \quad (77)$$

Assume that $|\bar{\lambda} + \bar{\mu}| < 0_1$ (0_1 : positive constant). Then, by using Taylor's expansion theorem, the following two inequalities can be obtained.

$$\exp[- x(1 - \frac{k}{n})\bar{\lambda}(1 - p_\beta) + \frac{k}{n} x \, \bar{\mu}(1 - p_\beta)]$$

$$\leq 1 - xp_\beta'((1 - \frac{k}{n})\bar{\lambda} - \frac{k}{n}\bar{\mu}) + x^2 p_\beta' 0_1^{\,2}[\exp(20_1 x)] \quad , \text{ where } p_\beta' = 1 - p_\beta(t) \quad (78)$$

$$\exp[x(1 - \frac{k}{n})\bar{\lambda}p_\beta - \frac{k}{n} x \, \bar{\mu} \, p_\beta]$$

$$\leq 1 + xp_\beta((1 - \frac{k}{n})\bar{\lambda} - \frac{k}{n}\bar{\mu}) + x^2 p_\beta 0_1^{\,2}[\exp(20_1 x)] \qquad (79)$$

From Lemma 2.1,

$$\sum_{k=0}^{n} D_{n,k}^i (1 - \frac{k}{n}) = 1 - \frac{1}{n}[n - (\sum_{j=1}^{n} c_j^i)] = \frac{1}{n}(\sum_{j=1}^{n} c_j^i) \qquad (80)$$

$$\sum_{k=0}^{n} D_{n,k}^i (\frac{k}{n}) = 1 - \frac{1}{n}(\sum_{j=1}^{n} c_j^i) \qquad (81)$$

Therefore, from (77) to (81), the following inequality can be obtained.

$$M(x,P) \leq 1 + \frac{xp_\beta p_\beta'}{n}(\bar{\lambda} + \bar{\mu})(\sum_{j=1}^{n} c_j^\delta - \sum_{j=1}^{n} c_j^\beta) + 2x^2 p_\beta p_\beta' 0_1^{\,2}[\exp(20_1 x)]$$

$$\leq 1 - xp_\beta p_\beta'[\frac{\tilde{\varepsilon}}{n}(\sum_{j=1}^{n} c_j^\delta - \sum_{j=1}^{n} c_j^\beta) + 2x0_1^{\,2}\exp(20_1 x)] \qquad (82)$$

$$(\sum_{j=1}^{n} c_j^\delta = \min_i \sum_{j=1}^{n} c_j^i \quad (i = 1,\ldots,r ; i \neq \beta) \qquad (68) \quad)$$

In the above inequality

$$\lim_{x \to 0} 2x\theta_1^2 \exp(2\theta_1 x) = 0 \qquad \text{and} \qquad \frac{\tilde{\varepsilon}}{n}(\sum_{j=1}^{n} C_j^\delta - \sum_{j=1}^{n} C_j^\beta) > 0 ,$$

Consequently, there exists some small positive number z such that

$$M(z,P) \leq 1$$

Hence, from (75)

$$E\{h_{z,\theta}(p_\beta'(t+1))/P(t)\} \leq h_{z,\theta}(p_\beta'(t))$$

for all t and P(t).

Q.E.D.

CHAPTER 3

LEARNING BEHAVIORS OF STOCHASTIC AUTOMATA UNDER
NONSTATIONARY MULTI-TEACHER ENVIRONMENT

3.1   Introduction

In the previous chapter, the GAE reinforcement scheme has been introduced
and it has been shown that this scheme has several desirable learning performances such as ε-optimality and absolute expediency in the general n-teacher
environment.   These properties have been obtained under the assumption that
the multi-teacher environment under consideration is stationary and P-model.
However, in many actual situations, such assumptions are found to be inadequate.

In this chapter, we consider learning behaviors of variable-structure stochastic automata operating in a nonstationary multi-teacher environment from
which stochastic automata receive responses having an arbitrary number between 0
and 1.   As a generalized form of the GAE reinforcement scheme given in the previous chapter, we propose the MGAE scheme and show that this scheme ensures
ε-optimality in the nonstationary multi-teacher environment of an S-model.

3.2 Learning Automaton Model under the Nonstationary Multi-Teacher Environment of S-model

In this section, we generalize the model given in Figure 2 and discuss the learning behaviors of the variable-structure stochastic automaton D in the nonstationary multi-teacher environment (NMT) as illustrated in Figure 9.

The stochastic automaton D is defined by the set $\{S,W,Y,g,P(t),T\}$. S is the set of inputs $(s_1^i,\ldots,s_n^i)$ where $s_j^i (j=1,\ldots,n)$ is the response from the jth teacher $R_j (j=1,\ldots,n)$ and the value of s_j^i can be an arbitrary number in the closed line segment $[0,1]$. (We are dealing with S-model multi-teacher environment.) In this model, the definitions of $W,Y,g,P(t)$, and T are the same as in the last chapter.

Assume now that the state w_i is chosen at time t. Then, the stochastic automaton D performs action y_i on the nonstationary multi-teacher environment (NMT). In response to y_i, the jth teacher R_j emits output s_j^i. We shall deal with the nonstationary multi-teacher environment in which s_j^i is a function of t and ω. ($\omega \in \Omega$; Ω is the basic space of the probability measure space (Ω,B,μ), and B is the smallest Borel field including $\bigcup_{t=0}^{\infty} F_t$, where $F_t = \sigma(P(0),\ldots,P(t)$, $S(0),\ldots,S(t)$) ($S(t)$ means the outputs from NMT at time t.)) From now on we shall often use the notation $s_j^i(t,\omega)$ to represent the input into the stochastic automaton D.

Depending upon the action y_i and the n responses $s_1^i(t,\omega),\ldots,s_n^i(t,\omega)$ from n teachers $R_1,\ldots,R_n$, the stochastic automaton D changes the probability vector $P(t)$ by the reinforcement scheme T.

The nonstationary multi-teacher environment (NMT) considered in this chapter has the property that the relation

$$\int_0^1 s\, dF_{\alpha,t}(s) + \frac{\delta}{n} < \int_0^1 s\, dF_{j,t}(s) \qquad (1)$$

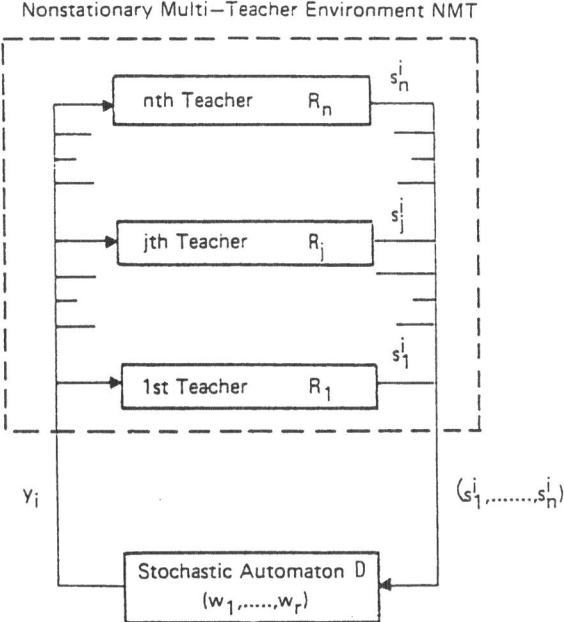

Figure 9   Stochastic automaton D operating in the nonstationary
            multi-teacher environment (NMT)

where $F_{i,t}(s)$ $(i=1,\ldots,r)$ is the distribution function of $\sum_{j=1}^{n} s_j^i(t,\omega)/n$, holds

for some state w_α, some $\delta > 0$, all time t, all j ($\neq \alpha$), and all ω ($\in \Omega$).

The objective of the stochastic automaton D is to reduce $E\{\sum_{j=1}^{n} s_j^i(t,\omega)\}$,

the expectation of the sum of the penalty strengths. Therefore, condition (1)

means that the αth action y_α is the best among r actions $y_1,\ldots,y_r$ since y_α

receives the least sum of the penalty strengths in the sense of mathematical

expectation.

In the previous chapter, we have defined several basic norms of the learn-

ing behaviors of stochastic automata under stationary multi-teacher environment.

By analogy from the definitions 2.4 and 2.5 introduced in the last chapter, we

can give the following definitions concerning learning norms of stochastic auto-

mata under nonstationary multi-teacher environment NMT satisfying the condition

(1).

Definition 3.1   The stochastic automaton D is said to be "optimal in NMT"

if

$$\lim_{t\to\infty} p_\alpha(t) = 1 \qquad \text{with probability 1} \qquad (2)$$

Definition 3.2   The stochastic automaton D is said to be "ε-optimal in NMT"

if the parameters of the reinforcement scheme can be chosen so that

$$\lim_{t\to\infty} E\{p_\alpha(t)\} \geq 1 - \varepsilon \qquad \text{for any } \varepsilon > 0 \qquad (3)$$

On the other hand, the extensions of the definitions 2.2 and 2.3 can not be

easily given.   Presumably, we need a different interpretation.

3.3 ε-Optimal Reinforcement Scheme under the Nonstationary Multi-Teacher Environment

The GAE reinforcement scheme has been introduced in the last chapter as

a class of learning algorithms of stochastic automata operating in a multi-teacher environment which emits 0 (reward) or 1 (penalty) responses. This scheme can not be applied to the S-model environment in which teachers emit arbitrary responses between 0 and 1.

In the following, let us propose the MGAE scheme which can be used for the S-model environment.

MGAE scheme :   Suppose that $y(t) = y_i$ and the responses from NMT are $(s_1^i, s_2^i, \ldots, s_n^i)$. Then,

$$p_i(t+1) = p_i(t) + [\frac{s_1^i + \ldots + s_n^i}{n}]\{\sum_{j \neq i}^{r} \phi_j(P(t))\} - [1 - \frac{s_1^i + \ldots + s_n^i}{n}]\{\sum_{j \neq i}^{r} \psi_j(P(t))\}$$

(4)

$$p_j(t+1) = p_j(t) - [\frac{s_1^i + \ldots + s_n^i}{n}]\{\phi_j(P(t))\} + [1 - \frac{s_1^i + \ldots + s_n^i}{n}]\{\psi_j(P(t))\}$$

    $(j \neq i)$

(5)

where ϕ_i, ψ_i $(i=1,\ldots,r)$ satisfy the following relations.

$$\frac{\phi_1(P(t))}{p_1(t)} = \ldots = \frac{\phi_r(P(t))}{p_r(t)} = \lambda(P(t))$$

(6)

$$\frac{\psi_1(P(t))}{p_1(t)} = \ldots = \frac{\psi_r(P(t))}{p_r(t)} = \mu(P(t))$$

(7)

$$p_j(t) + \psi_j(P(t)) > 0, \qquad p_i(t) + \sum_{j \neq i}^{r} \phi_j(P(t)) > 0$$

$$p_j(t) - \phi_j(P(t)) < 1 \qquad (j = 1,\ldots,r ; i = 1,\ldots,r)$$

(8)

Remark 3.1   The MGAE scheme is a generalized form of the GAE scheme introduced in the previous chapter. In (14) and (15) in chapter 2, let replace m by $(n - (s_1^i + \ldots + s_n^i))$. Then, we can easily obtain (4) and (5).

As to the learning performance of the MGAE reinforcement scheme, the following theorem can be obtained.

Theorem 3.1   Suppose that $\lambda(P(t))$ and $\mu(P(t))$ satisfy the assumptions of the theorem 2.2.   Then, the stochastic automaton D with the MGAE reinforcement scheme is ε-optimal under the nonstationary multi-teacher environment NMT satisfying condition (1).

In order to prove the above theorem, we need several important lemmas.

Lemma 3.1:   Suppose that all of the assumptions of the above theorem hold. Then, the MGAE reinforcement scheme has the following learning performance under the NMT environment satisfying condition (1):

$$E\{p_\alpha(t+1)/P(t)\} \geq p_\alpha(t) \qquad (9)$$

Proof: For notational convenience, let us abbreviate time t and probability vector P(t) as follows:

$$p_i \triangleq p_i(t), \quad \phi_i \triangleq \phi_i(P(t)), \quad \psi_i \triangleq \psi_i(P(t)), \quad \lambda \triangleq \lambda(P(t)), \quad \mu \triangleq \mu(P(t)).$$

$$(i = 1,\ldots,r)$$

Let $F_{i,t}(\xi)$ be the distribution function of

$$\frac{s_1^i(t,\omega) + \ldots + s_n^i(t,\omega)}{n} \qquad (i = 1,\ldots,r) \qquad (10)$$

Then, the conditional expectation $E\{p_\alpha(t+1)/P(t)\}$ can be calculated as follows:

$$E\{p_\alpha(t+1)/P(t)\} = p_\alpha \int_0^1 [p_\alpha + \xi(\sum_{j\neq\alpha}^{r} \phi_j) - (1-\xi)(\sum_{j\neq\alpha}^{r} \psi_j)] \, dF_{\alpha,t}(\xi)$$

$$+ \sum_{j\neq\alpha}^{r} p_j \int_0^1 [p_\alpha - \xi(\phi_\alpha) + (1-\xi)\psi_\alpha] \, dF_{j,t}(\xi)$$

$$= P_\alpha - P_\alpha (\sum_{j \neq \alpha}^{r} \psi_j) + P_\alpha [\sum_{j \neq \alpha}^{r} (\phi_j + \psi_j)][\int_0^1 \xi \, dF_{\alpha,t}(\xi)]$$

$$+ (1-P_\alpha)\psi_\alpha - (\phi_\alpha + \psi_\alpha)[\sum_{j \neq \alpha}^{r} P_j \int_0^1 \xi \, dF_{j,t}(\xi)] \qquad (11)$$

Let

$$C_k(t) = [\int_0^1 \xi \, dF_{k,t}(\xi)] \qquad (k = 1,\dots,r) \qquad (12)$$

Then, using the relations (6) and (7), the above equality can be represented as:

$$E\{p_\alpha(t+1)/P(t)\} = P_\alpha + P_\alpha(1 - P_\alpha)\{\lambda + \mu\}C_\alpha(t) - \{\lambda + \mu\}[\sum_{j \neq \alpha}^{r} p_j C_j(t)]P_\alpha \qquad (13)$$

From the definition of the distribution function $F_{k,t}(\xi)$, and from condition (1),

$$C_\alpha(t) + \frac{\delta}{n} < C_j(t) \qquad \text{for all } j \ (j \neq \alpha, \ 1 \leq j \leq r) \text{ and } \omega.$$

$$(14)$$

Let

$$C_\beta(t) = \min [C_{k_1}(t),\dots,C_{k_{r-1}}(t)] \qquad (k_1,\dots,k_{r-1} \neq \alpha) \qquad (15)$$

Then, from the relations (11) $\sim$ (15) and the assumptions of the lemma, we can get

$$E\{p_\alpha(t+1)/P(t)\} \geq P_\alpha(t) + \{\lambda+\mu\}(1-P_\alpha)P_\alpha[C_\alpha(t) - C_\beta(t)] \geq P_\alpha(t) \qquad (16)$$

$$[C_\alpha(t) - C_\beta(t) < 0 \quad \text{and} \quad \lambda+\mu < 0]$$

Q.E.D.

Lemma 3.2: Suppose that all of the assumptions of the theorem 3.1 hold.

Let

$$h_{x,\theta}(p) \triangleq \frac{\exp(xp/\theta) - 1}{\exp(x/\theta) - 1} \qquad (x > 0) \qquad (17)$$

$$p'_\alpha(t) \triangleq 1 - p_\alpha(t) \qquad (18)$$

Then, there exists some positive constant z which satisfies the inequality

$$E\{h_{z,\theta}(p'_\alpha(t+1))/P(t)\} \leq h_{z,\theta}(p'_\alpha(t)) \qquad \text{for all t and P(t).}$$

Proof: The conditional expectation $E\{h_{x,\theta}(p'_\alpha(t+1))/P(t)\}$ can be calculated as follows:

$$E\{h_{x,\theta}(p'_\alpha(t+1))/P(t)\}$$

$$= J[-1 + (\exp(xp'_\alpha/\theta))\{ p_\alpha \int_0^1 \exp[-xp'_\alpha(\xi\bar\lambda-(1-\xi)\bar\mu)] \, dF_{\alpha,t}(\xi)$$

$$+ \sum_{j\neq\alpha}^r p_j \int_0^1 \exp[xp_\alpha(\xi\bar\lambda-(1-\xi)\bar\mu) \, dF_{j,t}(\xi) \}] \qquad (19)$$

where
$$J = 1/(\exp(x/\theta) - 1), \qquad p'_\alpha = 1 - p_\alpha(t), \qquad \text{and} \quad p_\alpha = p_\alpha(t)$$

Assume that

$$| \bar\lambda + \bar\mu | < 0_2 \qquad (0_2 : \text{positive constant}) \qquad (20)$$

Then, the following two inequalities can be obtained:

$$\exp[-xp'_\alpha(\xi\bar\lambda-(1-\xi)\bar\mu) \leq 1 - xp'_\alpha(\xi\bar\lambda-(1-\xi)\bar\mu) + 20_2|\bar\lambda+\bar\mu|x^2 p'_\alpha[\exp(20_2x)] \qquad (21)$$

$$\exp[xp_\alpha(\xi\bar\lambda-(1-\xi)\bar\mu)] \leq 1 + xp_\alpha(\xi\bar\lambda-(1-\xi)\bar\mu) + 20_2|\bar\lambda+\bar\mu|x^2 p_\alpha[\exp(20_2x)] \qquad (22)$$

From (19),(21), and (22), we can get

$$E\{h_{x,\theta}(p'_\alpha(t+1))/P(t)\} \leq -J + J(\exp(xp'_\alpha/\theta))\{1 + p_\alpha x[\bar\lambda(\sum_{j\neq\alpha}^r p_j C_j(t) - p'_\alpha C_\alpha(t))$$

$$+ \bar\mu(\sum_{j\neq\alpha}^r p_j C_j(t) - p'_\alpha C_\alpha(t))$$

$$+ f_1(x,P)] \qquad (23)$$

where
$$f_1(x,P) = 4x0_2 p'_\alpha|\bar\lambda+\bar\mu| [\exp(20_2x)] \qquad (24)$$

From (14),

$$E\{h_{x,\theta}(p_\alpha'(t+1))/P(t)\} \leq -J + J(\exp(xp_\alpha'/\theta))[1 - p_\alpha p_\alpha' |\overline{\lambda+\mu}| x(\frac{\delta}{n} - 4xO_2\exp(2O_2 x))]$$

(25)

In the above inequality,

$$\lim_{x\to 0} 4xO_2\exp(2O_2 x) = 0, \qquad p_\alpha p_\alpha' |\overline{\lambda+\mu}| \geq 0, \quad \text{and} \quad \frac{\delta}{n} \text{ is a positive constant.}$$

Hence, there exists some positive constant z which satisfies the inequality

$$E\{h_{z,\theta}(p_\alpha'(t+1))/P(t)\} \leq h_{z,\theta}(p_\alpha'(t)) \qquad \text{for all } t \text{ and } P(t).$$

Q.E.D.

Proof of Theorem 3.1:

We have obtained Lemma 3.1 and Lemma 3.2. Therefore, using the same procedure as done in the previous chapter, we can easily prove Theorem 3.1.

Q.E.D.

Remark 3.2   We have discussed the learning behaviors of the MGAE scheme under the nonstationary multi-teacher environment of S-model satisfying the condition (1). Therefore, the theorem obtained above can also hold true under the P-model nonstationary multi-teacher environment. This can be stated as follows:

Theorem 3.1'   Assume that the nonstationary multi-teacher environment satisfies:

$$C_\alpha(t,\omega) + \delta < C_j(t,\omega)$$

(26)

, where $C_i(t,\omega)$ $(i=1,\ldots,r)$ is the probability of the penalty response to the ith output y_i of the stochastic automaton at time t, holds for some state w_α, some $\delta > 0$, all time t, all j ($j \neq \alpha$), and all ω ($\in \Omega$).

Then, the MGAE (GAE) reinforcement scheme (satisfying the assumptions of the Theorem 2.2) ensures ε-optimality under the nonstationary multi-teacher environment satisfying (26).

3.4 Computer Simulation Results

In the last section, we proposed the MGAE reinforcement scheme as an extension of the GAE scheme and showed that it ensures ε-optimality under the nonstationary multi-teacher environment of S-model satisfying the condition (1). In this section, we present several computer simulation results of the learning behaviors of the stochastic automaton under the nonstationary multi-teacher environments.

Example 1 :   In this example, we simulate by computer the learning behaviors of the two-state stochastic automaton with the GL_{R-I} scheme (See Remark 2.12) under the nonstationary two-teacher environment of P-model ($R_1(C_1^1(t,\omega),C_1^2(t,\omega))$, $R_2(C_2^1(t,\omega),C_2^2(t,\omega))$).   Here, $C_i^j(t,\omega)$ ($i,j = 1,2$) are the uniformly distributed random variables having the following ranges:

$$\{\tfrac{1}{10}\sin(\tfrac{t}{100}\pi) + 0.25\} - 0.1 \ \leq\ C_1^1(t,\omega)\ \leq\ \{\tfrac{1}{10}\sin(\tfrac{t}{100}\pi) + 0.25\} + 0.1$$

$$\{-\tfrac{1}{10}\sin(\tfrac{t}{200}\pi)+0.85\} - 0.1 \ \leq\ C_1^2(t,\omega)\ \leq\ \{-\tfrac{1}{10}\sin(\tfrac{t}{200}\pi)+0.85\} + 0.1$$

$$\{\tfrac{1}{10}\sin(\tfrac{t}{50}\pi) + 0.35\} - 0.05 \ \leq\ C_2^1(t,\omega)\ \leq\ \{\tfrac{1}{10}\sin(\tfrac{t}{50}\pi) + 0.35\} + 0.05$$

$$\{-\tfrac{1}{5}\sin(\tfrac{t}{50}\pi) + 0.60\} - 0.05 \ \leq\ C_2^2(t,\omega)\ \leq\ \{-\tfrac{1}{5}\sin(\tfrac{t}{50}\pi) + 0.60\} + 0.05 \qquad (27)$$

In Figure 10, brief sketches of $C_i^j(t,\omega)$ (i,j = 1,2) and sum of the two penalty probabilities are given. The learning behaviors of the GL_{R-I} scheme under the above two-teacher environment are simulated by computer and given in Figure 11.

Example 2 : In this example, we consider the nonstationary two-teacher environment ($R_1(C_1^1(t,\omega),C_1^2(t,\omega))$, $R_2(C_2^1(t,\omega),C_2^2(t,\omega))$) which is slightly different from that in Example 1. That is, $C_i^j(t,\omega)$ (i,j = 1,2) are also uniformly distributed random variables having the following ranges:

$$\{\frac{1}{10} \sin(\frac{t}{100}\P) + 0.25\} - 0.1 \leq C_1^1(t,\omega) \leq \{\frac{1}{10} \sin(\frac{t}{100}\P) + 0.25\} + 0.1$$

$$\{-\frac{1}{10} \sin(\frac{t}{100}\P)+0.55\} - 0.1 \leq C_1^2(t,\omega) \leq \{-\frac{1}{10} \sin(\frac{t}{200}\P)+0.55\} + 0.1 \quad .$$

$$\{\frac{1}{10} \sin(\frac{t}{50}\P) + 0.35\} - 0.05 \leq C_2^1(t,\omega) \leq \{\frac{1}{10} \sin(\frac{t}{50}\P) + 0.35\} + 0.05$$

$$\{-\frac{1}{10} \sin(\frac{t}{50}\P) + 0.5\} - 0.05 \leq C_2^2(t,\omega) \leq \{-\frac{1}{10} \sin(\frac{t}{50}\P) + 0.5\} + 0.05 \quad (28)$$

The learning behaviors of the GL_{R-I} scheme under the above two-teacher environment are simulated by computer and given in Figure 12.

66

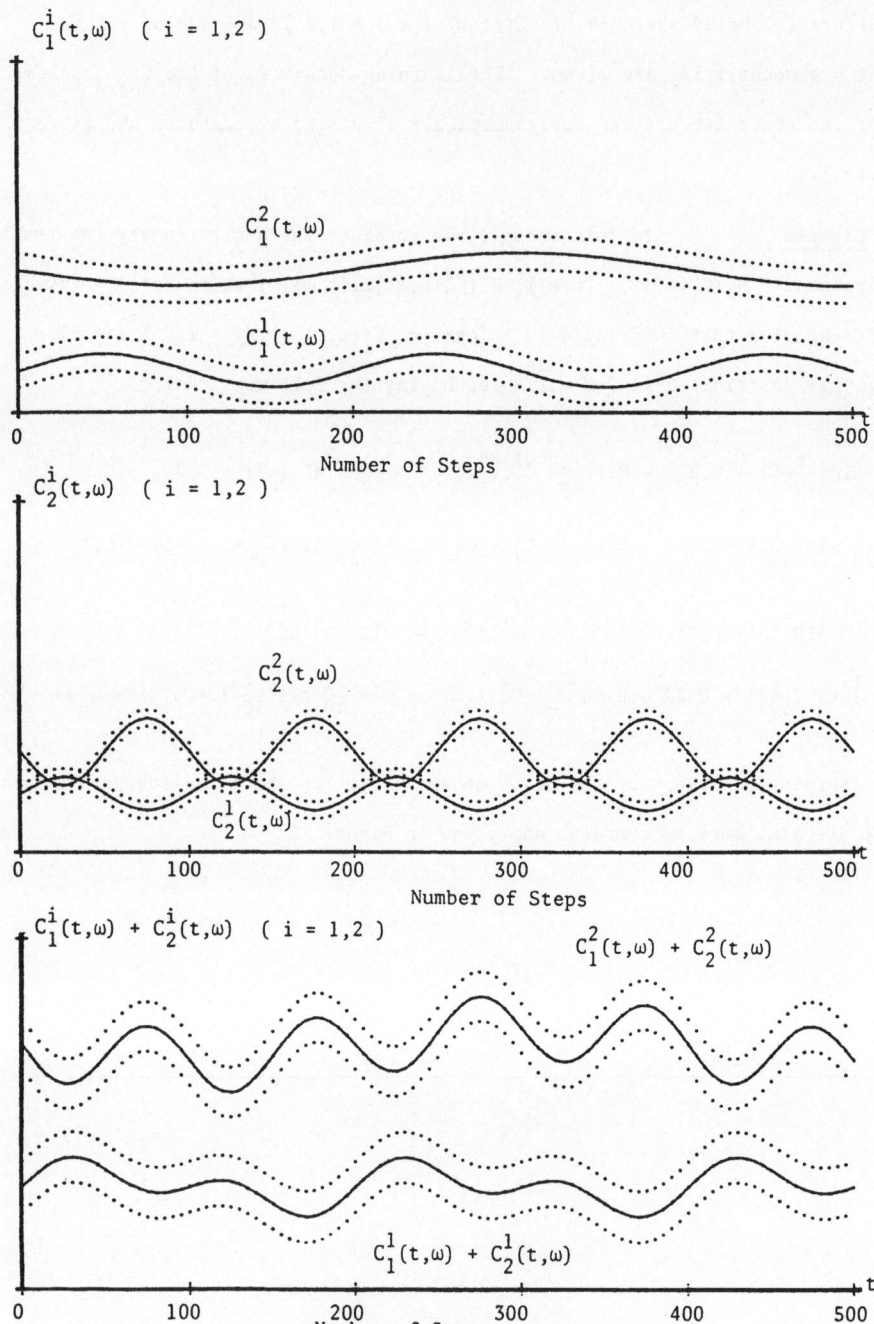

Figure 10   Sum of the penalty probabilities in Example 1

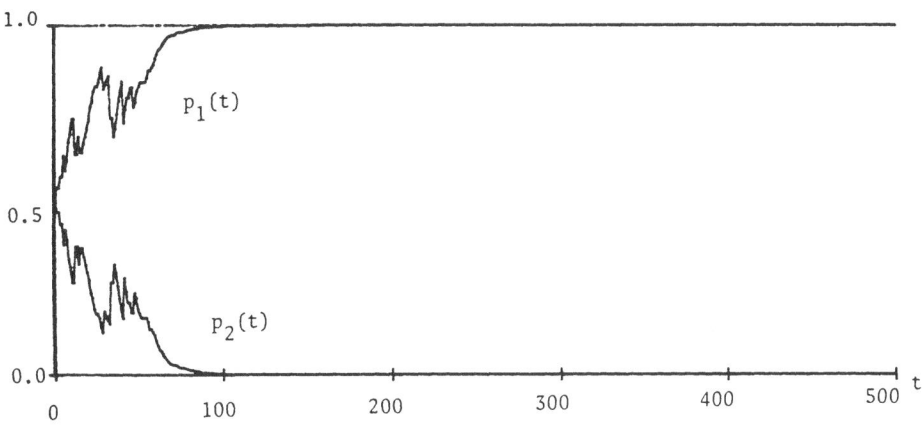

Figure 11   Changes in probability $p_i(t)$ ($i = 1,2$) in

Example 1

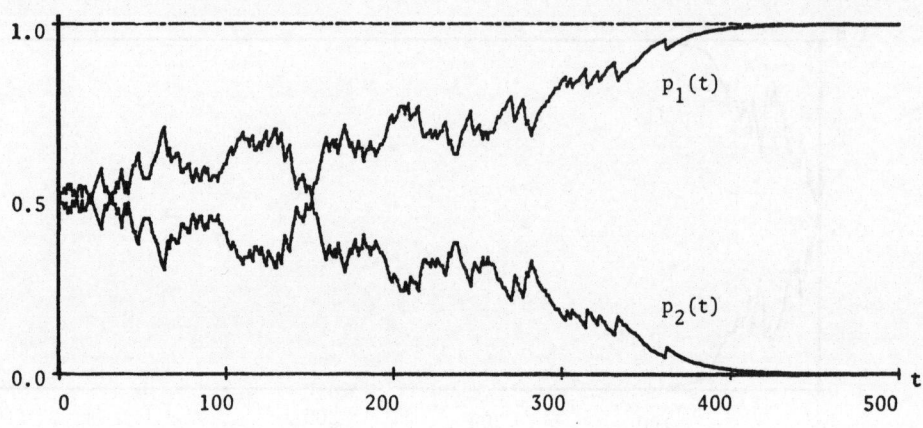

Figure 12   Changes in probability $p_i(t)$ ($i = 1,2$) in

Example 2

Remark 3.3　　　　The nonstationary two-teacher environment in the first example satisfies the condition (1)'. ($\alpha = 1$)　　　The computer simulation result shown in Figure 11 confirms our theoretical study. (The state probability $p_1(t)$ increases with time t and tends to 1.0.)

Remark 3.4　　　　In the second example, the penalty probabilities $c_i^j(t,\omega)$ are slightly different from those in the first example.　　While $c_1^1(t,\omega)$ + $c_2^1(t,\omega)$ is smaller than $c_1^2(t,\omega)$ + $c_2^2(t,\omega)$ in almost all part of the product space $T \times \Omega$ ($t \in T$, $\omega \in \Omega$), the condition (1) does not hold true.　　(In some region of the product space $T \times \Omega$, $c_1^1(t,\omega)$ + $c_2^1(t,\omega)$ is larger than $c_1^2(t,\omega)$ + $c_2^2(t,\omega)$.)　　Therefore, the state probability $p_1(t)$ does not increase so fast as in the first example.

Remark 3.5　　　　In this section, we have only given several computer simulation results of the learning behaviors of the 2-state stochastic automaton under the nonstationary 2-teacher environment of P-model.　　In order to discuss the learning performance of the MGAE scheme under the nonstationary multiteacher environment, we should test it on various examples including S-model environments.

3.5　Comments and Concluding Remarks

The learning behaviors of a variable-structure stochastic automaton operating in the stationary random environment $R(C_1,\ldots,C_r)$ have been extensively studied by many researchers.

However, compared with the great number of studies related to the behavior of learning automaton in a stationary environment, only a few and specialized results have been obtained concerning those in a nonstationary environment. Following the research by Chandrasekaran and Shen [C2], Narendra and Viswanathan [N1] considered periodically changing nonstationary random environment with unknown period. They proposed the two level system of stochastic automata. Baba and Sawaragi [B1] considered the nonstationary random environment which has the property that $C_\alpha(t,\omega) + \delta < C_j(t,\omega)$ holds for some α, some $\delta > 0$, and all j ($\neq \alpha$), and all ω; ω is a point of a basic ω-space Ω. They showed that the L_{R-I} scheme ensures ε-optimality under the above environment. Recently, Srikantakumar and Narendra [S10] studied the learning behaviors of stochastic automata under the following nonstationary random environment: (i) $C_i(P(n))$ ($i=1,\ldots,r$; $n=0,\ldots$) are continuous functions of p_i.

(ii) $\dfrac{\partial C_i}{\partial p_i} > 0$ for any i.

(iii) $\dfrac{\partial C_i}{\partial p_i} \gg \dfrac{\partial C_i}{\partial p_j}$ for any i,j. ($i \neq j$)

This work has a very interesting application in the area of telephone network routing.

In this chapter, we have discussed the learning behavior of stochastic automata under the nonstationary multi-teacher environment (NMT) in which penalty strengths are functions of t and ω, where t represents time and ω is a point of the basic ω-space Ω. It has been proved that the MGAE reinforcement scheme, which is an extended form of the GAE reinforcement scheme introduced in chapter 2, ensures ε-optimality under the nonstationary multi-teacher environment (NMT) which satisfies the condition (1).

Generally speaking, learning behaviors of stochastic automata under nonstationary environments are difficult to analyze, compared with those under stationary environments. However, in order to make stochastic automata theory more attractive, we must investigate learning behaviors of stochastic automata under various nonstationary environments which could find important practical applications.

CHAPTER 4

APPLICATION TO NOISE-CORRUPTED,
MULTI-OBJECTIVE PROBLEM

4.1   Introduction

Theoretical studies about the learning behaviors of stochastic automata have been advanced by many researchers.   The applications of these studies to the practical problems have also been reported.   For example, Waltz and Fu [W1] tried to use stochastic automata in an unknown control system, Chandrasekaran and Shen [C3] applied stochastic automata to two person zero-sum games, Riordon [R3] used stochastic automata as the learning controllers of a control system having unknown Markov process, Shapiro and Narendra [S6] utilized stochastic automata for the parameter self-optimization problem with unknown performance criteria.   Recently, the routing of messages in communication networks has been found to be a quite promising application area of stochastic automata. ([M3], [S10], and etc.)

In this chapter, we consider a parameter self-optimization problem with noise-corrupted, multi-objective functions as an application of learning behaviors of stochastic automata operating in an unknown nonstationary multi-teacher environment.

4.2   Statement of the Problem

Suppose that the $J_1(\alpha),\ldots,$ and $J_n(\alpha)$ are unknown objective functions of a parameter $\alpha \in [\alpha_1,\ldots,\alpha_r]$ except that they are bounded $(-M \le J_1(\alpha),\ldots,J_n(\alpha) \le M)$. It is assumed that measurements of $J_i(\alpha)$ $(i=1,\ldots,n)$ can be obtained only from the noise-corrupted observations.

$$g_i(\alpha,\xi) = J_i(\alpha) + \xi_i \qquad (i = 1,\ldots,n) \tag{1}$$

, where ξ_i is an additive Gaussian white noise with zero mean and variance ρ.

Here, $J_i(\alpha)$ is assumed to have unique maximum $J_i(\alpha_{\beta_i})$:

$$J_i(\alpha_{\beta_i}) = \max [J_i(\alpha_1),\ldots,J_i(\alpha_r)] \tag{2}$$

Each objective function $J_i(\alpha)$ has the claim to be maximized. $(i = 1,\ldots,n)$. However, generally, the relation $\alpha_{\beta_1} = \alpha_{\beta_2} = \ldots = \alpha_{\beta_n}$ does not hold. This is one of the most difficult points of multi-objective optimization porblems.

4.3   An Application of the Stochastic Automaton to the Noise-Corrupted, Multi-Objective Problem

The learning behaviors of stochastic automata having been studied in the last chapter can be used to find an appropriate parameter in this problem. Let us try to identify the ith action y_i of stochastic automaton D with the ith parameter value α_i $(i=1,\ldots,r)$. Choosing the ith parameter α_i at time t corresponds to D producing the output y_i at time t. For simplicity, we consider the stochastic automaton D under P-model environment.

Let k_t^j be a measurement of $g_j(\alpha, \xi_j)$ at time t. Further, let $\bar{k}_t^j$ (t = 0,1,... ; j = 1,...,n) be defined as

$$\bar{k}_0^j = \begin{cases} k_0^j & \text{if } -M \le k_0^j \le M \\ M & \text{if } k_0^j > M \\ -M & \text{if } k_0^j < -M \end{cases}$$

$$\bar{k}_t^j = \begin{cases} \frac{1}{t+1}(t\cdot\bar{k}_{t-1}^j + k_t^j) & \text{if } -M \le k_t^j \le M \\ \frac{1}{t+1}(t\cdot\bar{k}_{t-1}^j + M) & \text{if } k_t^j > M \\ \frac{1}{t+1}(t\cdot\bar{k}_{t-1}^j - M) & \text{if } k_t^j < -M \end{cases} \tag{3}$$

Using these values, we define reward and penalty as follows.

Suppose that $y(t) = \alpha_i$ (i = 1,...,r). If $k_t^j \ge \bar{k}_{t-1}^j$, then the stochastic automaton D receives reward response $s_j^i = 0$ from the jth teacher R_j (j = 1,...,n). (This means that the jth noise-corrupted, objective function $J_j(\alpha)$ gives an affirmative answer to the parameter α_i.) On the contrary, if $k_t^j < \bar{k}_{t-1}^j$, then the stochastic automaton D receives penalty response $s_j^i = 1$ from the jth teacher R_j (j = 1,...,n). The stochastic automaton changes the state vector P(t) to P(t+1) by the n responses ($s_1^i,...$ ..,s_n^i) which it has received from the n teachers $R_1,...,$ and R_n.

Now, let us consider the learning behavior of D. If the parameter α_i is selected at time t, D receives penalty from the jth teacher R_j with the probability

$$\mu(g_j(\alpha_i, \xi_j(t)) < \bar{k}_{t-1}^j)$$

From (1),

$$\mu(\ g_j(\alpha_i,\xi_j(t)) < \overline{k}^j_{t-1}\ ) \ = \ \mu(\ \xi_j(t) < \overline{k}^j_{t-1} - J_j(\alpha_i)\ )$$

$$= \ P_{\xi_j}(\ \xi_j(t) < \overline{k}^j_{t-1} - J_j(\alpha_i)\ ) \tag{4}$$

($P_{\xi_j}(\cdot)$ is the distribution function of ξ_j ($j = 1,\dots,n$).)

Since $J_j(\alpha)$ is assumed to have unique maximum $J_j(\alpha_{\beta_j})$,

$$\mu(\ \xi_j(t) < \overline{k}^j_{t-1} - J_j(\alpha_{\beta_j})\ ) \ < \ \mu(\ \xi_j(t) < \overline{k}^j_{t-1} - J_j(\alpha)\ ) \tag{5}$$

for all $\overline{k}^j_{t-1}$ and all α ($\alpha_{\beta_j} \neq \alpha$) ($j = 1,\dots,n$)

(See Figure 13.)

Let

$$c^i_j(t,\omega) \ = \ \mu(\ g_j(\alpha_i,\xi_j(t)) < \overline{k}^j_{t-1}\ ) \tag{6}$$

The reason why we use the notation $c^i_j(t,\omega)$ is to represent the probability that stochastic automaton D receives penalty response from the jth teacher when it selects the ith parameter α_i at time t.   (Here, $\omega \in \Omega$, Ω being the supporting set of the probability measure space (Ω, B, μ). $F_t = \sigma(P(0),\dots, P(t),\overline{k}^1_0,\dots,\overline{k}^n_0,\overline{k}^1_1,\dots,\overline{k}^n_1,\dots,\overline{k}^n_t)$ ($\sigma(P(0),\dots,\overline{k}^n_t)$ is the smallest Borel field of ω-sets with respect to which $P(0),\dots,$ and $\overline{k}^n_t$ are all measurable.) B is the smallest Borel field which contains $\bigcup_{t=0}^{\infty} F_t$. μ is the probability measure which satisfies $\mu(\Omega) = 1$.)

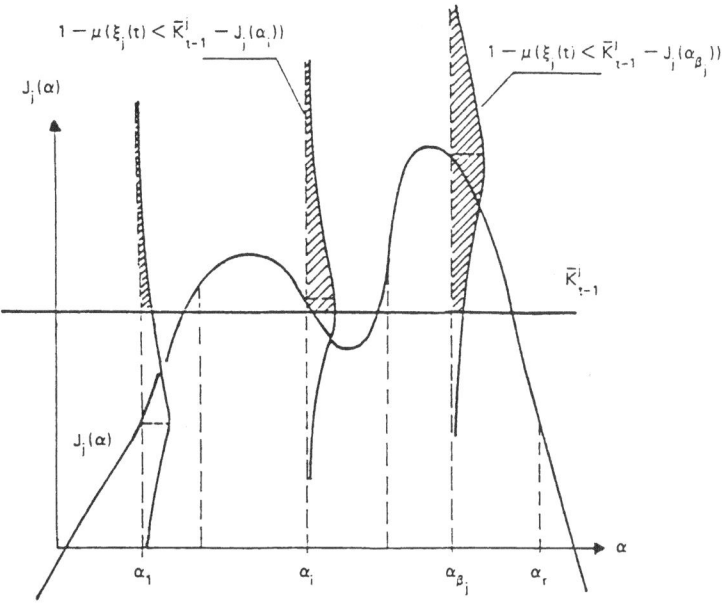

Figure 13   The value of $1 - \mu(\xi_j(t) < \bar{k}^j_{t-1} - J_j(\alpha_{\beta_j}))$

Therefore, it follows from (2),(4),(5), and (6) that

$$c_j^{\beta_j}(t,\omega) + \delta_j < c_j^i(t,\omega) \tag{7}$$

for all t, all i (i = 1,...,r ; i $\neq$ β_j), all $\omega \in \Omega$, and some positive ·number δ_j (j = 1,...,n)

If the strict condition

$$\alpha_\beta^* = \alpha_{\beta_1} = \ldots = \alpha_{\beta_n} \tag{8}$$

holds, then it can be easily derived from (7) that

$$c_1^{\beta^*}(t,\omega) + \ldots + c_n^{\beta^*}(t,\omega) + \delta < c_1^i(t,\omega) + \ldots + c_n^i(t,\omega) \tag{9}$$

$$(\delta = \delta_1 + \ldots + \delta_n)$$

for all t, all i (i = 1,...,r ; i $\neq$ β^*), all $\omega \in \Omega$, and the positive number δ.

Therefore, using the theoretical results obtained in the last chapter, we can prove that

$$\lim_{\theta \to 0} \lim_{t \to \infty} E\{ p_{\beta*}(t) \} = 1 \tag{10}$$

is ensured by the MGAE reinforcement scheme.

However, generally, the condition (8) does not hold. Even in such a case, the MGAE reinforcement scheme can successfully find appropriate parameters. In the following section, we give several computer simulation results which indicate the nice learning property of the MGAE reinforcement scheme.

Remark 4.1   Let us briefly touch upon the learning property of the MGAE reinforcement scheme in the general case.

First of all, let us give a definition of the Pareto-optimal parameters in

the noise-corrupted, multi-objective problem by extending the original definition
of Pareto-optimal solutions.

Definition 4.1 α_{i_k} $(k=1,\ldots,q)$ is said to be Pareto-optimal parameter if
there does not exist such a parameter that satisfies

i) $J_i(\alpha) \geq J_i(\alpha_{i_k})$ for all i ($1 \leq i \leq n$)   and

ii) $J_p(\alpha) \neq J_p(\alpha_{i_k})$ for some p ($1 \leq p \leq n$)

For the later use, we also give the following definition.

Definition 4.2   Let $\alpha_{i_1},\ldots,\alpha_{i_q}$ be the Pareto-optimal parameters.   Then,
a parameter α_γ is said to be "not completely inferior to one of the Pareto-
optimal parameters" if there exists some parameter α_{i_s} ($1 \leq s \leq q$)   and
$J_k(\alpha_\gamma) > J_k(\alpha_{i_s})$ for some k ($1 \leq k \leq n$).

Let $p_{\delta_1}(t)$ be the sum of the state probabilities corresponding to the
Pareto-optimal parameters.   Further, let $p_{\delta_2}(t)$ be the sum of the state
probabilities whose corresponding parameters are not completely inferior to one
of the Pareto-optimal parameters.

Let
$$p_\delta(t) \overset{\Delta}{=} p_{\delta_1}(t) + p_{\delta_2}(t)$$

Then, it can be easily derived that

$$\lim_{\theta \to 0} \lim_{t \to \infty} E\{ p_\delta(t) \} = 1$$

4.4 Computer Simulation Results

In this section, we present three computer simulation results which illustrate learning performance of the MGAE reinforcement scheme under the noise-corrupted, multi-objective problem.

Example 1 :   Let us consider the following noise-corrupted, multi-objective problem:

a) $n = 2$; $J_1(\alpha) = - (\alpha - 2)^2 + 10$

$J_2(\alpha) = - 2\alpha + 12$ (See Figure 14)

b)   Both ξ_1 and ξ_2 are the Gaussian white noises with zero mean and variance 0.1.

In the above problem, we consider the stochastic automaton having 5 actions $y_1,\ldots,y_5$ whose ith action y_i corresponds to the parameter $\alpha = i$.   The changes in the probabilities $p_i(t)$ $(i=1,\ldots,5)$ are shown in Figure 15.

Example 2 :   Let us consider the following noise-corrupted, multi-objective problem:

a) $n = 3$; $J_1(x_1,x_2) = 100(x_1^2 - x_2)^2 + (1 - x_2)^2$

$J_2(x_1,x_2) = [1 + (x_1+x_2+1)^2(19-14x_1+3x_1^2-14x_2+6x_1x_2+3x_2^2)]$

$[30 + (2x_1-3x_2)^2(18-32x_1+12x_1^2+48x_2-36x_1x_2+27x_2^2)]$

$J_3(x_1,x_2) = 4x_1^2 - 2.1x_1^4 + \frac{1}{3}x_1^6 + x_1x_2 - 4x_2^2 + 4x_2^4$

b) ξ_1, ξ_2, and ξ_3 are the Gaussian white noises with zero mean and variance 0.1, zero mean and variance 2.0, and zero mean and variance 0.5, respectively.

In the above problem, we consider the stochastic automaton having 9 actions $y_1, \ldots, y_9$ whose ith action y_i corresponds to the parameter as shown in Figure 16. The changes in the probabilities $p_i(t)$ $(i=1, \ldots, 9)$ are presented in Figure 17.

Example 3 : Let us consider the same noise-corrupted, multi-objective problem as the Example 2. In this example, we consider the stochastic automaton having 25 states whose ith state corresponding to the parameter as shown in Figure 18. The changes in the probabilities $p_i(t)$ $(i=1, \ldots, 25)$ are shown in the figures 19 to 21.

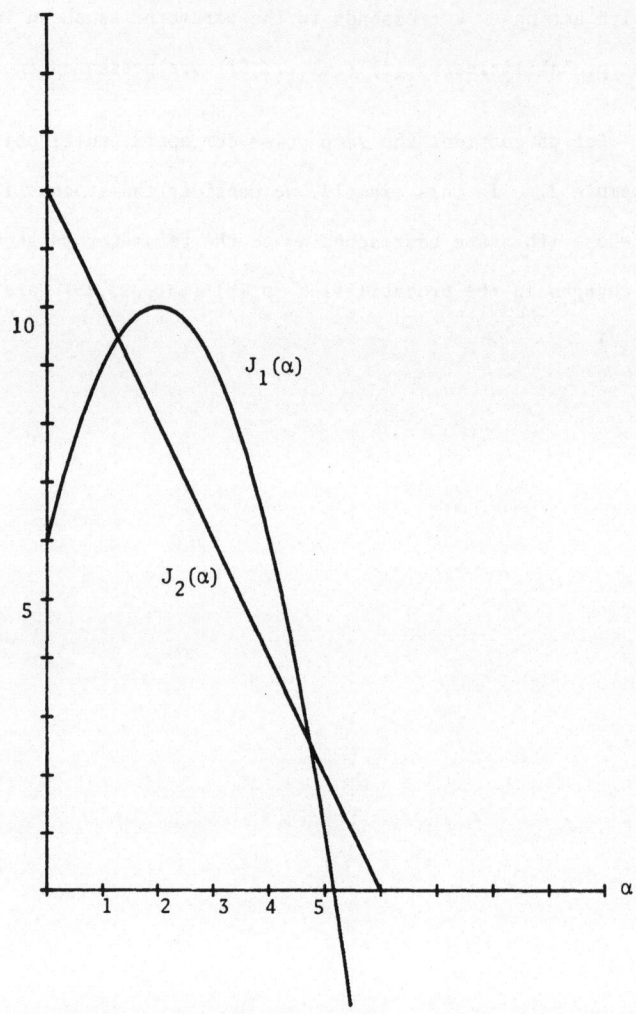

Figure 14   Two Objective Functions $J_1(\alpha)$ and $J_2(\alpha)$

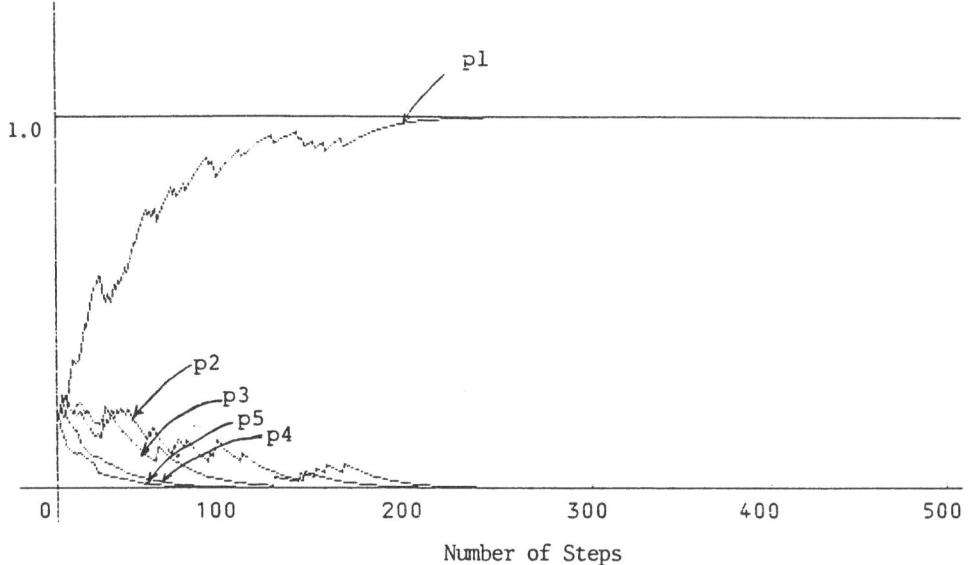

Figure 15   Changes in the probabilities $p_i(t)$ (i=1,...,5) in

Example 1.

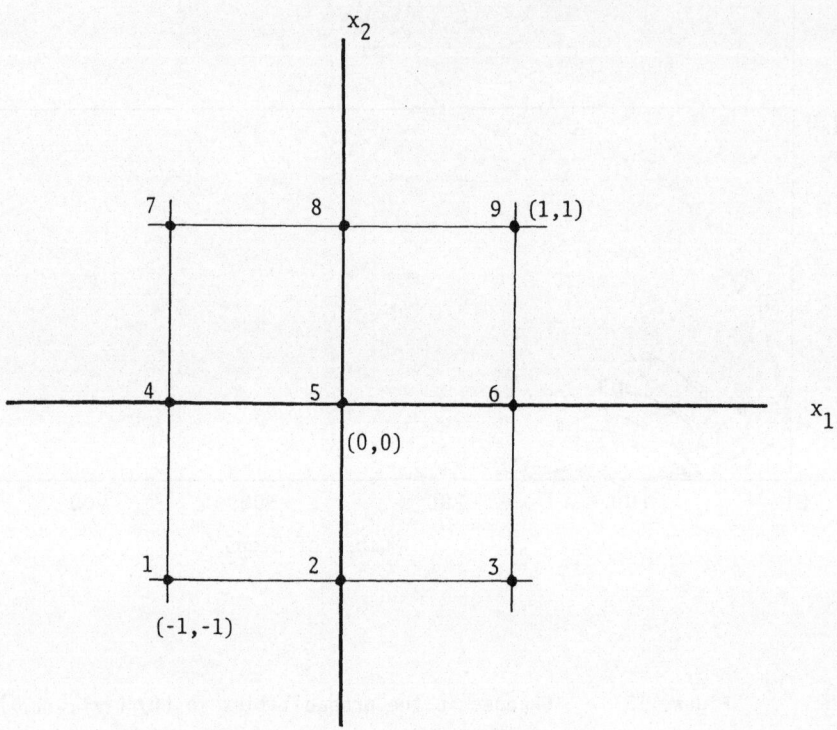

Figure 16   Parameters in Example 2

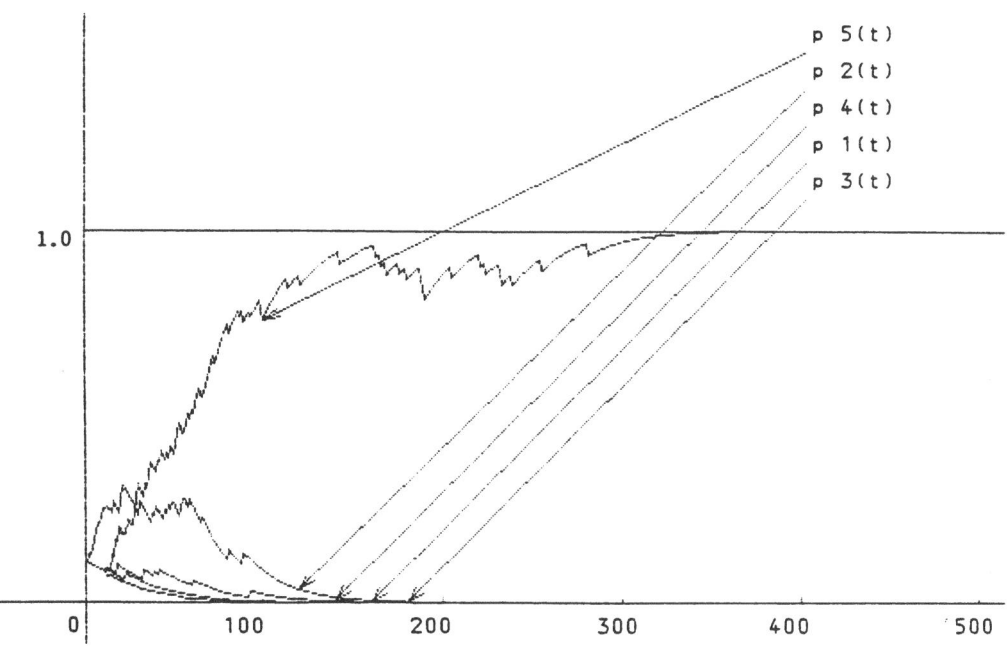

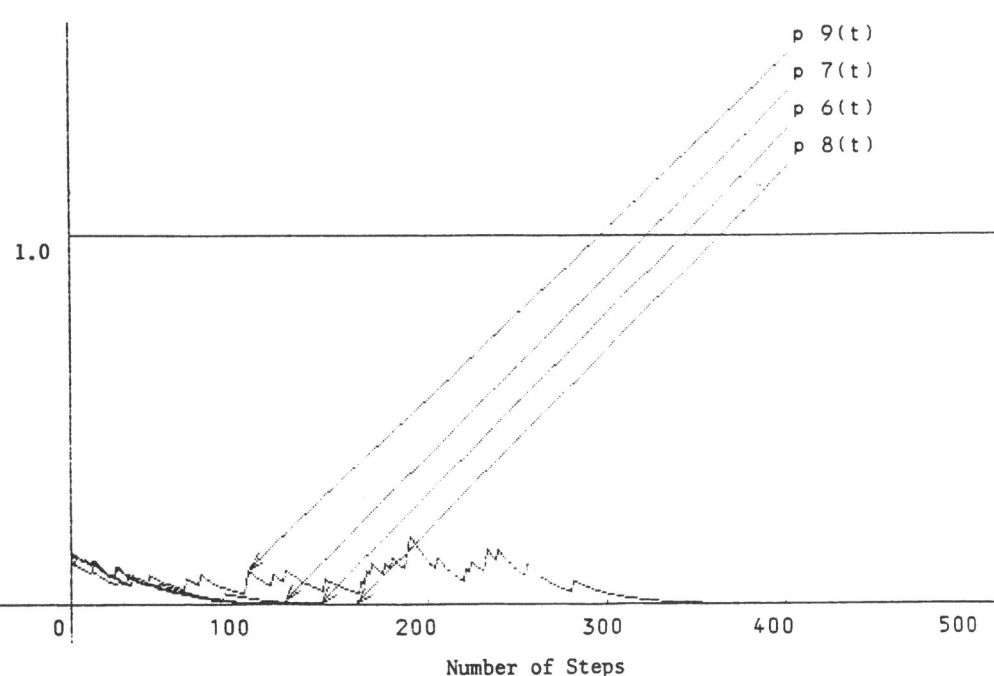

Figure 17   Changes in the probabilities $p_i(t)$ (i=1,...,9) in

Example 2

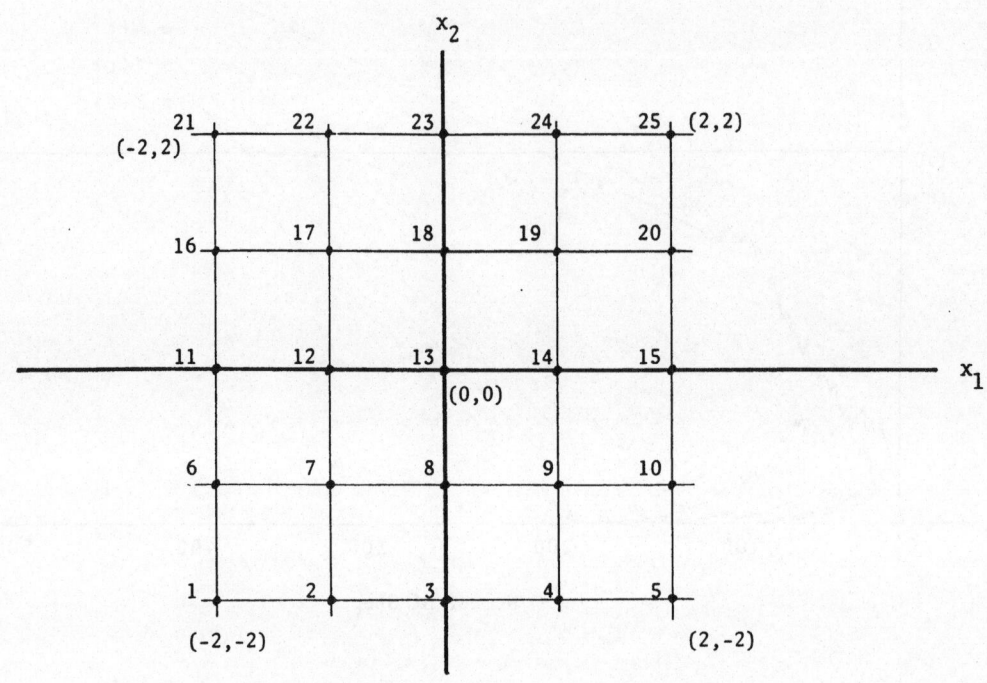

Figure 18   Parameters in Example 3

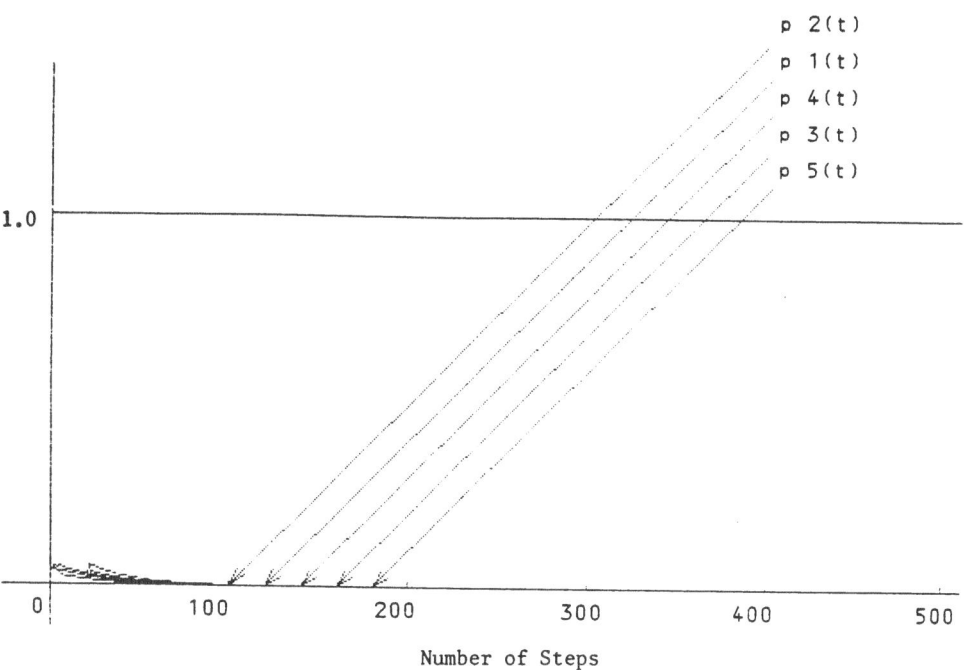

Number of Steps

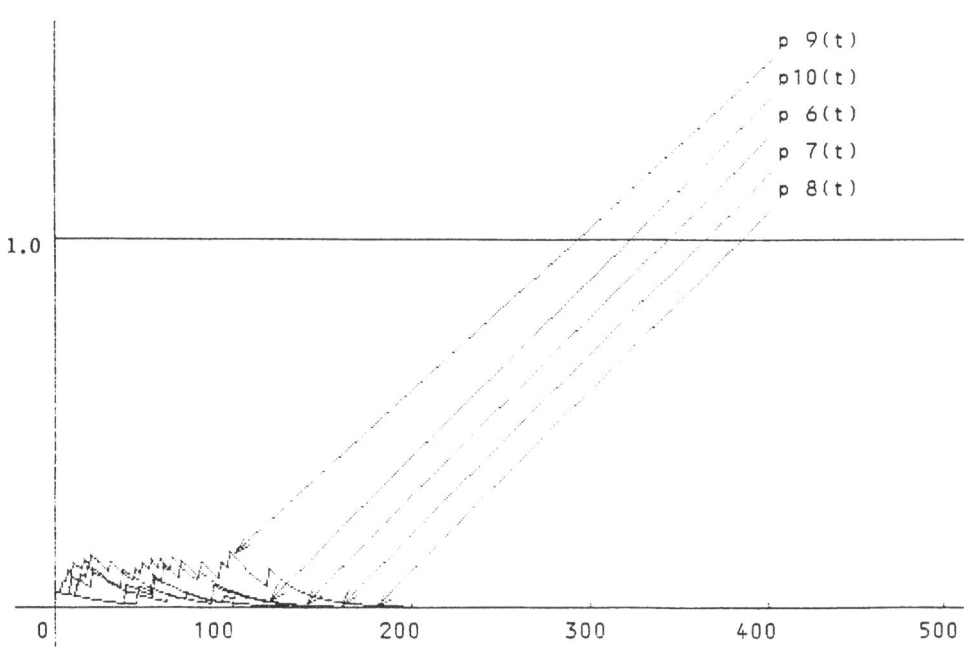

Figure 19   Changes in the probabilities $p_i(t)$ (i=1,...,10) in

Example 3

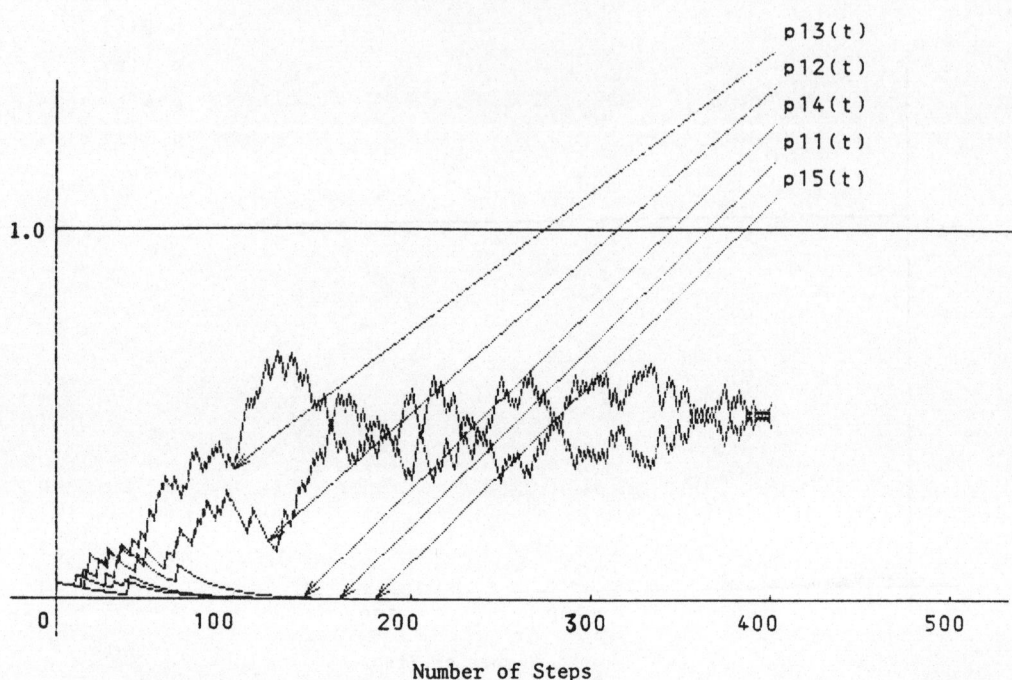

p13(t)
p12(t)
p14(t)
p11(t)
p15(t)

Number of Steps

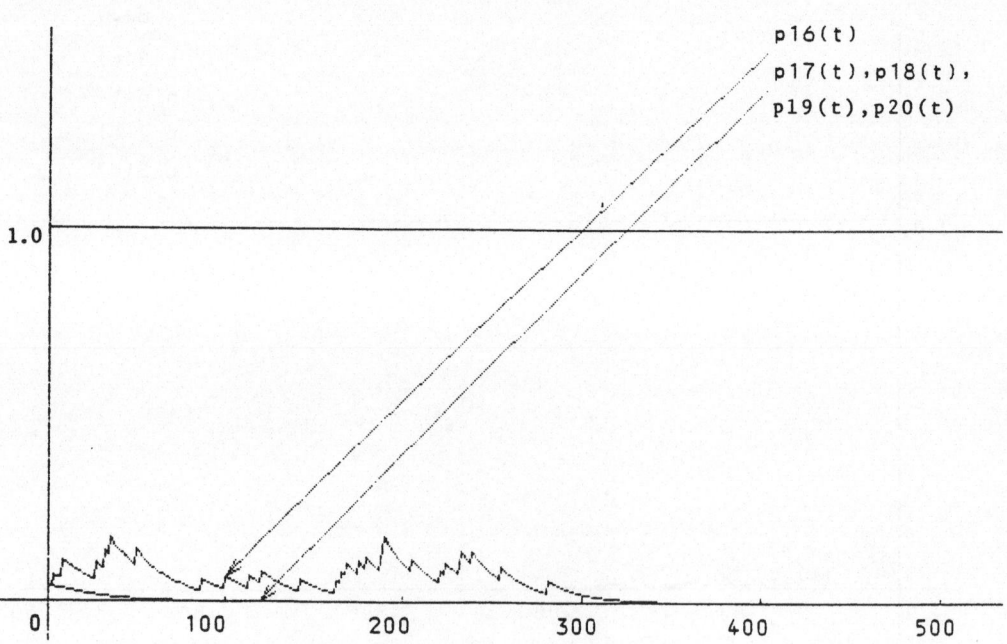

p16(t)
p17(t),p18(t),
p19(t),p20(t)

Figure 20   Changes in the probabilities $p_i(t)$ (i=11,...,20) in

Example 3

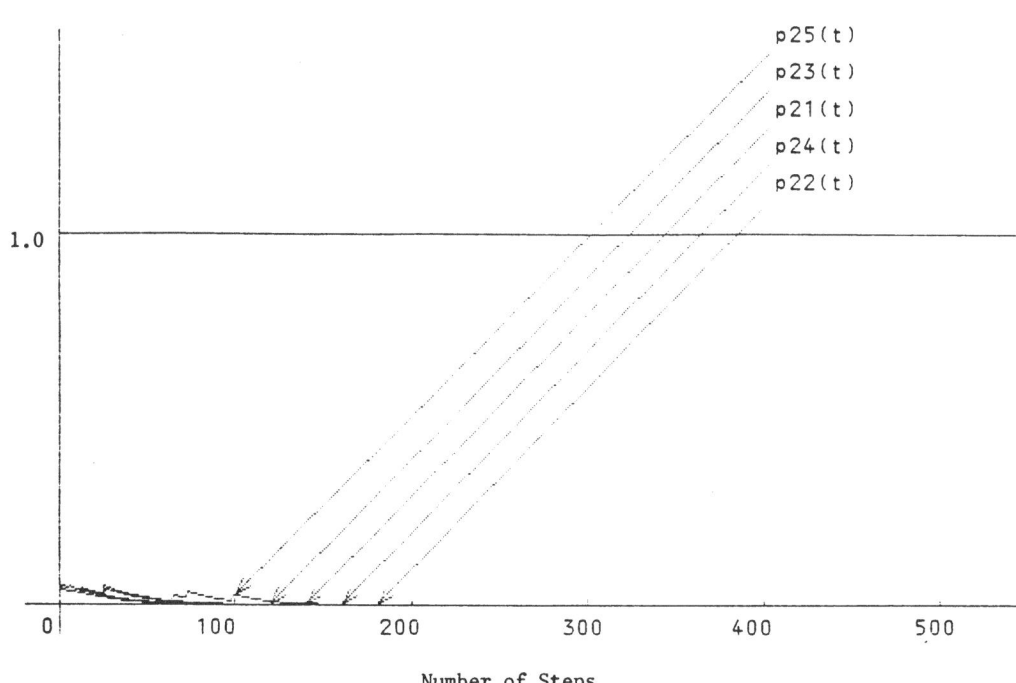

Figure 21   Changes in the probabilities $p_i(t)$ $(i=21,\ldots,25)$ in

Example 3

All of the above three examples do not satisfy the strict condition (8).
Consequently, there are several Pareto-optimal parameters in each of the three
examples. (Example 1: α_1 and α_2 ;   Example 2: α_2, α_5, and α_9 ;   Example 3:
α_8, α_{13}, and α_{19})   In example 1 (example 2), the state probability of one
of the Pareto-optimal parameters increases and tends to 1.0 with time t.   In
example 3, sum of the state probabilities $p_{12}(t)$ and $p_{13}(t)$ increases and
tends to 1.0 with time t.   (α_{12} is a Pareto-optimal parameter and α_{13} is a
parameter not completely inferior to one of the Pareto-optimal parameters.)
This means that the above computer simulation results confirm our theoretical
study. ((11))

4.5   Comments and Concluding Remarks

The ordinary (not noise-corrupted) multi-objective problem has been
studied extensively by many researchers and it has reached some level of
maturity. (See [H1], [G8], and etc.)   On the other hand, the noise-corrupted
multi-objective problem has not been studied so much.   There have not been found
any powerful approaches to this problem.

In this chapter, we have proposed the stochastic automaton's approach to
the parameter self-optimization problem with noise-corrupted, multi-objective
functions and showed that this problem can be reduced to the learning behaviors
of stochastic automaton operating in the nonstationary multi-teacher environment
(NMT).   We have obtained the following theoretical result: "If the strict condition (8) holds, the MGAE reinforcement scheme ensures ε-optimality
$(\lim_{\theta \to 0} \lim_{t \to \infty} E\{ p_{*\beta}(t) \} = 1)$"   We have also briefly mentioned to the general case
in which the strict condition (8) does not hold.   However, the result is a
rather weak one.

In our approach, stochastic automaton elects some of the appropriate parameters only from the finite candidates of the parameters ($\alpha_1, \ldots, \alpha_r$). It does not investigate all of the parameters which may be able to become candidates. Therefore, this might be one of the most important defects in the stochastic automaton's approach to the noise-corrupted, multi-objective problem. Therefore, an active future research to correct this defect is needed.

The hierarchical structure automata could be suggested as one of the most promising approaches to mitigate this difficulty.

AN APPLICATION OF THE HIERARCHICAL STRUCTURE AUTOMATA TO THE COOPERATIVE GAME WITH INCOMPLETE INFORMATION

5.1   Introduction

In the previous chapter, we considered a noise-corrupted, multi-objective problem as an application of stochastic automaton operating in a nonstationary multi-teacher environment.   Although this application appeared to be quite promising, it has the drawback that only limited number of state can be investigated.

In this chapter, we consider an application of learning automata to the game with incomplete information.   Game theory has been studied quite extensively by many authors and applied to various fields.   In an original setup of the general game theory, it is assumed that all players participated in the game are able to get sufficient informations concerning the game.   Recently, considerable attention has been directed to the analysis of the games with incomplete information. [H2] [H4],[K7],[P4]   In this chapter, we consider the cooperative game in which a player can only get informations about his gain and loss in each of the repeated game.   It is shown that the hierarchical structure automata are extremely helpful for finding an appropriate strategy in the game with incomplete information.   In the appendix of this chapter, we generalize the hierarchical structure automata model introduced by Ramakrishnan [R1] and consider the learning behaviors of the hierarchical structure stochastic automata operating in the general multi-teacher environments.

5.2   Statement of the Problem

Assume that three players A,B,C are participating in the coalitional game. Each of the three players wants to make coalition with one of the other two players in each of the repeated games.   If he cannot make any coalition, his payoff becomes zero. ($v(A) = v(B) = v(C) = 0$)   If any coalition has been set up between two players, then they must decide how to divide the value of the coalition.

Let us explain this more clearly.   The player A must decide the player with whom he has coalition in each of the repeated games.   If the coalition with player B has been set up, he must negotiate with B for the division of v(AB) (value of the coalition between A and B).   Let us assume that there are m kinds of divisions. (If A has succeeded in making coalition with C, he must also choose one division from m alternatives.)   If B does not agree with A's proposal (the way of division of v(AB)), A cannot get any return.

It is assumed that the player A knows neither the value of coalitions, nor the probabilities with which the other player disagrees with his proposal.   He can only get information about his return in each of the repeated games.

In the following, it will be shown that the hierarchical structure stochastic automata are very useful for the above problem.

5.3   Hierarchical Structure Stochastic Automata

In this section, we will briefly explain the learning mechanism of the hierarchical structure automata [R1] under the unknown environment.   The learning mechanism of the hierarchical structure automata in which each automaton in the hierarchy is able to elicit a response from the environment is described in Figure 22.

Ramakrishnan [R1] proposed the following reinforcement algorithm:

Ist Level

Assume that the j_1th action α_{j_1} is selected by the first level automaton A at time t. Then,

$$P_{j_1}(t+1) = P_{j_1}(t) + L_1(t)(1 - P_{j_1}(t))$$

$$P_{i_1}(t+1) = P_{i_1}(t)(1 - L_1(t)) \qquad (i_1 \neq j_1 \; ; \; i_1 = 1,\ldots,r)$$

Nth Level $\qquad (N = 2,\ldots,N)$

Assume that the actions α_{j_1}, $\alpha_{j_1 j_2}$,..., and $\alpha_{j_1 j_2 \ldots j_{N-1} j_N}$ are selected by the automata $A,\ldots,A_{j_1 \ldots j_{N-1}}$ at time t.

Then,

$$P_{j_1 j_2 \ldots j_N}(t+1) = P_{j_1 \ldots j_N}(t) + L_N(t)(1 - P_{j_1 \ldots j_N}(t))$$

$$P_{j_1 \ldots j_{N-1} i_N}(t+1) = P_{j_1 \ldots j_{N-1} i_N}(t)(1 - L_N(t))$$

$$(i_N \neq j_N \; ; \; i_N = 1,\ldots,r)$$

All other state probabilities at the Nth level of other automata remain unchanged.

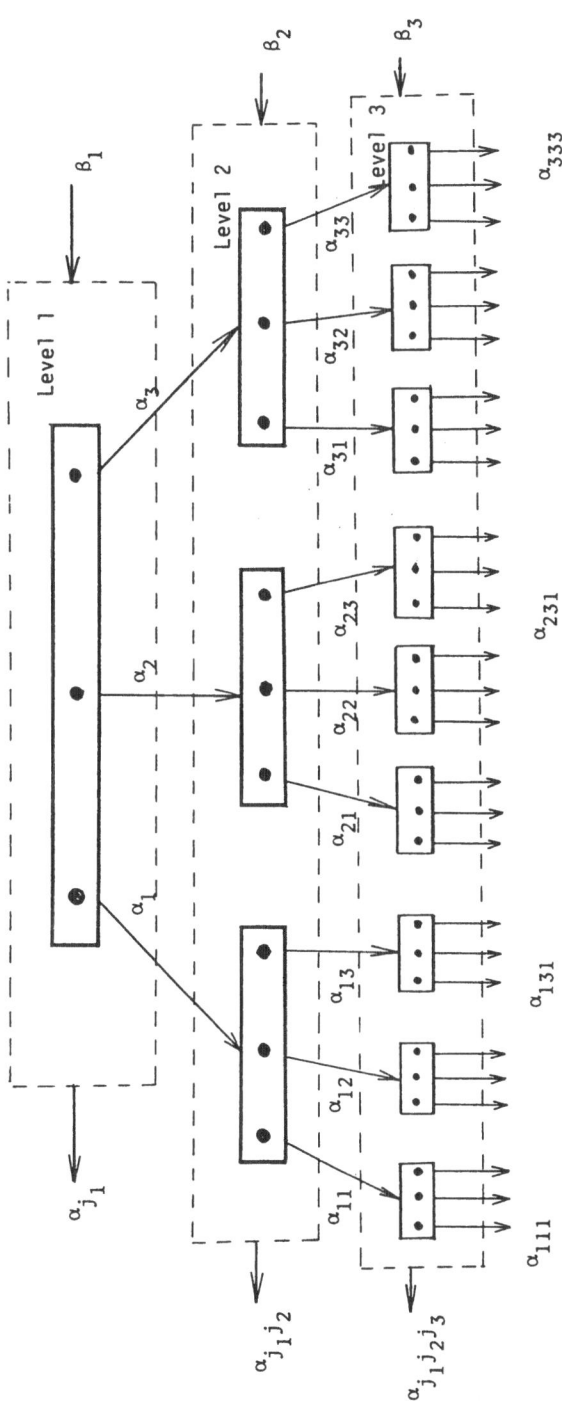

Figure 22   An Example of the Hierarchical Structure Stochastic Automata

He derived the following theorem:

If the conditions

$$L_1(t) = \sum_{s=1}^{N} \eta_s(t)\lambda_s ,$$

$$\vdots$$

$$L_n(t) = \frac{\sum\limits_{s=n}^{N} \eta_s(t)\lambda_s}{p_{j_1}(t+1)p_{j_2}(t+1)...p_{j_{n-1}}(t+1)} ,$$

$$(n = 2,3,...,N)$$

$$\eta_s(t) = 1 - \beta_s(t) \quad (s = 1,...,N), \quad \text{where} \quad \beta_s(t)$$

is the environmental response, and

$$0 < \sum_{s=1}^{N} \lambda_s < 1$$

are satisfied, the above algorithm is absolutely expedient.

5.4 An Application of the Hierarchical Structure Automata to the Cooperative Game

In this section, we show how the hierarchical structure automata can be applied to the above cooperative game. As we have explained in the previous section, the player A must decide the partner who makes coalition with him and negotiate for the division of the coalitional value. In the above setup of the cooperative game, let us consider the following two-level hierarchical structure automata.

Ist Level

Let $\quad \overline{V_t(A,B)}$:   The average return of the player A until time t by A's making coalition with B.

$\quad\quad \overline{G(t-1)}$:   The average return of the player A until time (t-1).

$\quad\quad \overline{V_t(A,C)}$:   The average return of the player A until time t by A's making coalition with C.

The input to the first level automaton A from the environment is determined as follows.

1)   A makes coalition with B

$\quad\quad$ If $\overline{G(t-1)} > \overline{V_t(A,B)}$, $\quad \beta_1(t) = 1.$

$\quad\quad$ If $\overline{G(t-1)} \leq \overline{V_t(A,B)}$, $\quad \beta_1(t) = 0.$

2)   A makes coalition with C

$\quad\quad$ If $\overline{G(t-1)} > \overline{V_t(A,C)}$, $\quad \beta_1(t) = 1.$

$\quad\quad$ If $\overline{G(t-1)} \leq \overline{V_t(A,C)}$, $\quad \beta_1(t) = 0.$

2nd Level

Let $\quad v_t(A,B)$:   The return of the player A at time t by A's negotiation with player B

$\quad\quad v_t(A,C)$:   The return of the player A at time t by A's negotiation with player C

1)   Assume that A has made coalition with B and decided how to divide V(AB).

If $\overline{V_{t-1}(A,B)} > v_t(A,B)$, $\beta_2(t) = 1$

If $\overline{V_{t-1}(A,B)} \leq v_t(A,B)$, $\beta_2(t) = 0$

2) Assume that A has made coalition with C and decided how to divide V(AC).

If $\overline{V_{t-1}(A,C)} > v_t(A,C)$, $\beta_2(t) = 1$

If $\overline{V_{t-1}(A,C)} \leq v_t(A,C)$, $\beta_2(t) = 0$

Initially, all of the state probabilities of hierarchical structure automata are set equal to $1/r$. (Ist level: $1/2$, 2nd level: $1/m$) Depending upon the environmental responses, the hierarchical structure automata change their state probabilities by the reinforcement scheme introduced in the previous section.

5.5   Computer Simulation Results

In this section, we present two computer simulation results.

Example 1:   In this example, v(AB) is a Gaussian random variable which has mean value 5.9 and variance 1.0.   v(AC) is also a Gaussian random variable which has mean value 3.2 and variance 1.0.   When A makes coalition with B, A proposes one of the following 5 alternatives concerning the division of v(AB).

1) $X_1 = v(AB)/3$ 2) $X_2 = v(AB)/2$ 3) $X_3 = v(AB)/4$

4) $X_4 = 2v(AB)/3$ 5) $X_5 = 3v(AB)/4$

If B agrees with A's proposal, B receives X_k (k=1,...,5) and A gets $v(AB) - X_k$.

When A makes coalition with C, A proposes one of the following alternatives concerning the division of v(AC).

1) $\overline{X}_1 = v(AC)/3$ 2) $\overline{X}_2 = v(AC)/2$ 3) $\overline{X}_3 = v(AC)/4$

4) $\overline{X}_4 = 2v(AC)/3$ 5) $\overline{X}_5 = 3v(AC)/4$

B (or C) does not necessarily agree with A's proposal.   If B (or C) rejects A's proposal, then A gets no return.

The probability that B (C) rejects A's proposal is shown in the followings:

1) The probability that B rejects A's proposal:

   0.75, 0.18, 0.88, 0.10, 0.07.

2) The probability that C rejects A's proposal:

   0.78, 0.23, 0.92, 0.15, 0.11.

If C agrees with A's proposal, C receives $\overline{X}_k$ (k=1,...,5) and A gets $v(AC) - \overline{X}_k$.

Figure 23 and Figure 24 present the computer simulation results of the above example.   In Figure 23, both the probability that A makes coalition with B in the first level and the probability that A makes coalition with B and selects the 2nd proposal in the 2nd level are gradually increased.   In Figure 24, the average return of A increases with time t.   These two computer simulation results indicate that the hierarchical structure automata method is quite effective for the cooperative game under discussion.   (It can be easily shown that A can get the highest return by making coalition with B and proposes the 2nd division.)

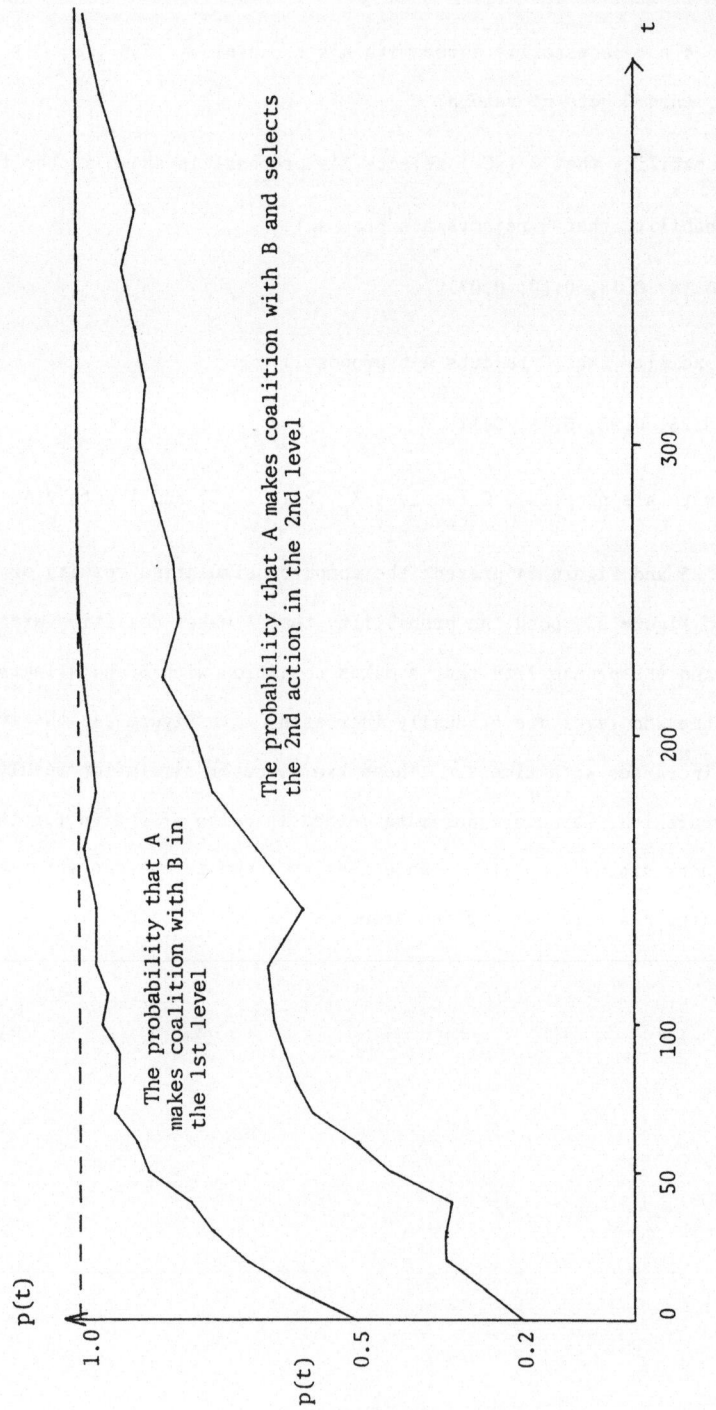

Figure 23   Changes in the action probabilities in Example 1

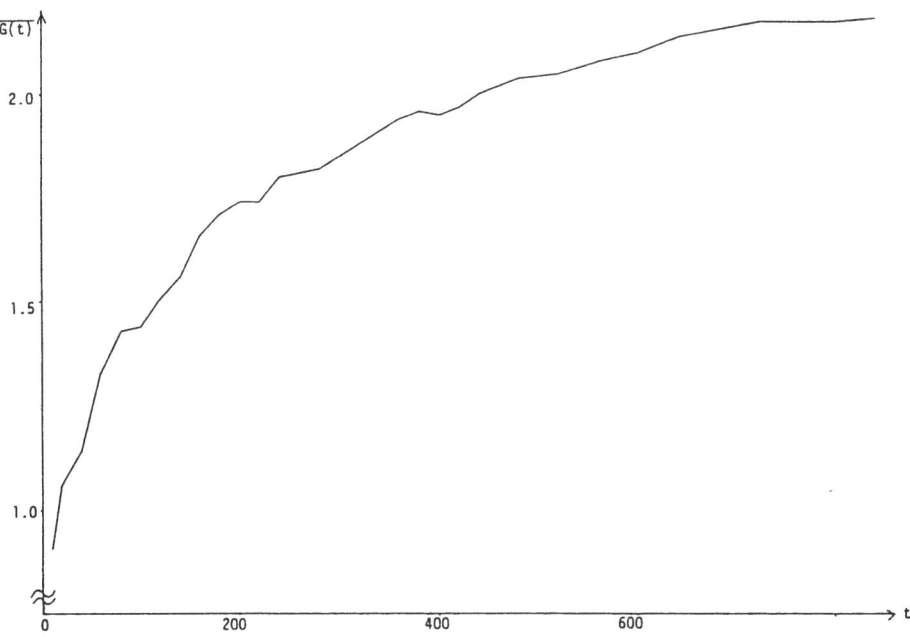

Figure 24   Changes in $\overline{G(t)}$ in Example 1

Example 2:   v(AB) is a Gaussian random variable which has mean value
5.5 and variance 1.0.   v(AC) is also a Gaussian random variable which has mean
value 8.0 and variance 1.0.   When A makes coalition with B, A proposes one of the
following 5 alternatives concerning the division of v(AB).

1) $X_1 = 4v(AB)/5$ 2) $X_2 = v(AB)/3$ 3) $X_3 = v(AB)/5$

4) $X_4 = v(AB)/2$ 5) $X_5 = 2v(AB)/3$

The probability that B rejects A's proposal is:

 0.08, 0.78, 0.95, 0.40, 0.28.

When A makes coalition with C, A proposes one of the 5 alternatives $\overline{X}_k$
1) $4v(AC)/5$ 2) $v(AC)/3$ 3) $v(AC)/5$ 4) $v(AC)/2$ 5) $2v(AC)/3$

The probability that C rejects A's proposal is:

 0.03, 0.65, 0.85, 0.15, 0.10.

Figure 25 and Figure 26 present the computer simulation results of Example 2.
In Figure 25, both the probability that A makes coalition with C in the first level
and the probability that A makes coalition with C and selects the 4th proposal in
the 2nd level are gradually increased.   In Figure 26, the average return of A
increases with time t.   These computer simulation results also demonstrate the
effectiveness of the hierarchical structure automata method.

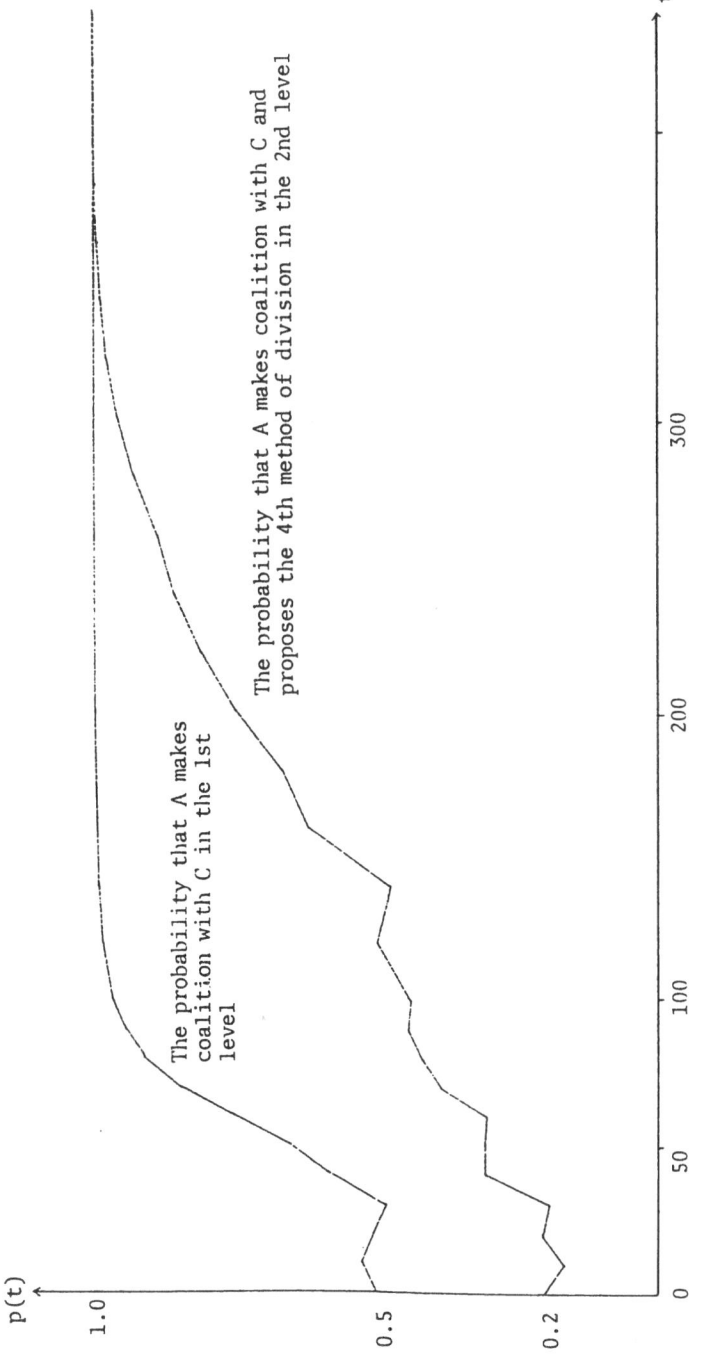

The probability that A makes coalition with C in the 1st level

The probability that A makes coalition with C and proposes the 4th method of division in the 2nd level

Figure 25   Changes in the action probabilities in Example 2

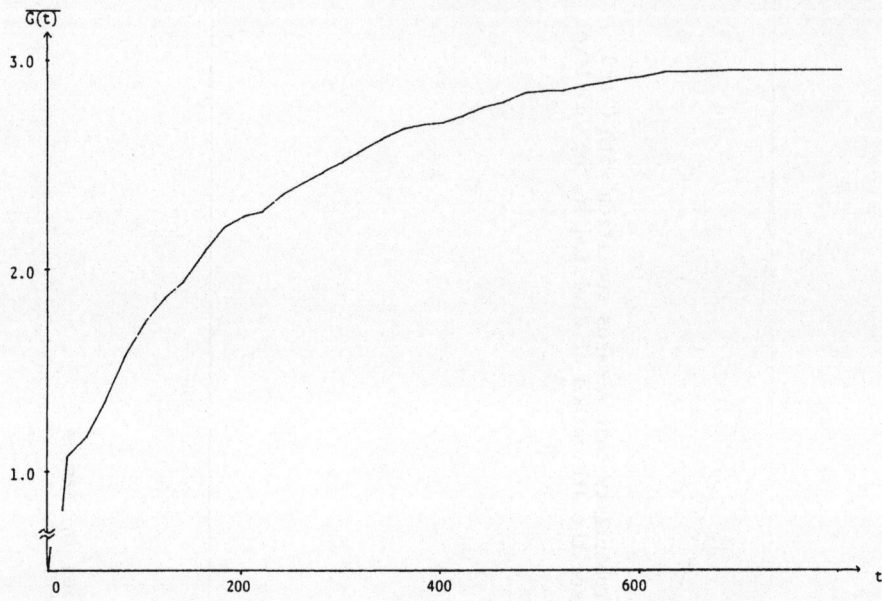

Figure 26   Changes in $\overline{G(t)}$ in Example 2

5.6   Comments and Concluding Remarks

The concept of the hierarchical structure automata was introduced recently by Thathachar and Ramakrishnan [T2].   This work was followed by Ramakrishnan [R1]. He formulated the more general hierarchical structure automata system in which each automaton in the hierarchy is able to elicit a response from the environment in each level.

In this chapter, we applied this hierarchical structure automata system to a simple coalitional game.   It was shown that using the hierarchical structure automata is an efficient method to find a partner to cooperate with and decide the division of the coalitional value.

Since many living systems follow hierarchical plans and our daily decision makings are often done in a hierarchical fashion, the hierarchical structure automata will surely find various interesting application areas.

In the last chapter, we considered stochastic automata approach to the noise-corrupted, multi-objective problem.   Although it was shown that stochastic automaton with an appropriate reinforcement scheme might be an effective tool in this problem, the main drawback is that stochastic automaton elects some of the parameters only from the limited number of candidates.   The concept of the hierarchical structure automata operating in the general multi-teacher environment could be successfully utilized in order to mitigate this drawback.

In the following appendix, we will briefly discuss the learning behaviors of the hierarchical structure stochastic automata operating in the general stationary multi-teacher environments.

5.7   Appendix - - - Learning Behaviors of the Hierarchical Structure

Stochastic Automata Operating in the General Multi-Teacher Environments

In this chapter, we applied the concept of the hierarchical structure automata to a cooperative game.   Since this concept could be applied in various areas,
it would become one of the most important promising tools.

The learning behaviors of the hierarchical structure automata operating in
the single teacher environment was considered by Thathachar and Ramakrishnan [T2]
and Ramakrishnan [R1].   They showed that absolute expediency can be assured under
some conditions.

In this appendix, we generalize their model and consider the learning behaviors of the hierarchical structure stochastic automata operating in the general
stationary multi-teacher environments.   Figure 27 describes the learning mechanism of the generalized model.   Almost all of the symbols used in the Figure 27
are the same as those used in the Figure 22, and so, we don't introduce them again.
We only mention to the new symbols.

For simplicity, let us consider the P-model environments.

Let β_s^i be the environmental response from the ith teacher at the s level.

($i = 1,\ldots,r_s$; $s = 1,\ldots,N$ | $\beta_s^i = 0$: reward response, $\beta_s^i = 1$: penalty

   response)

By the simple extension of the work done by Ramakrishnan [R1], we are able to get
the following learning algorithm.

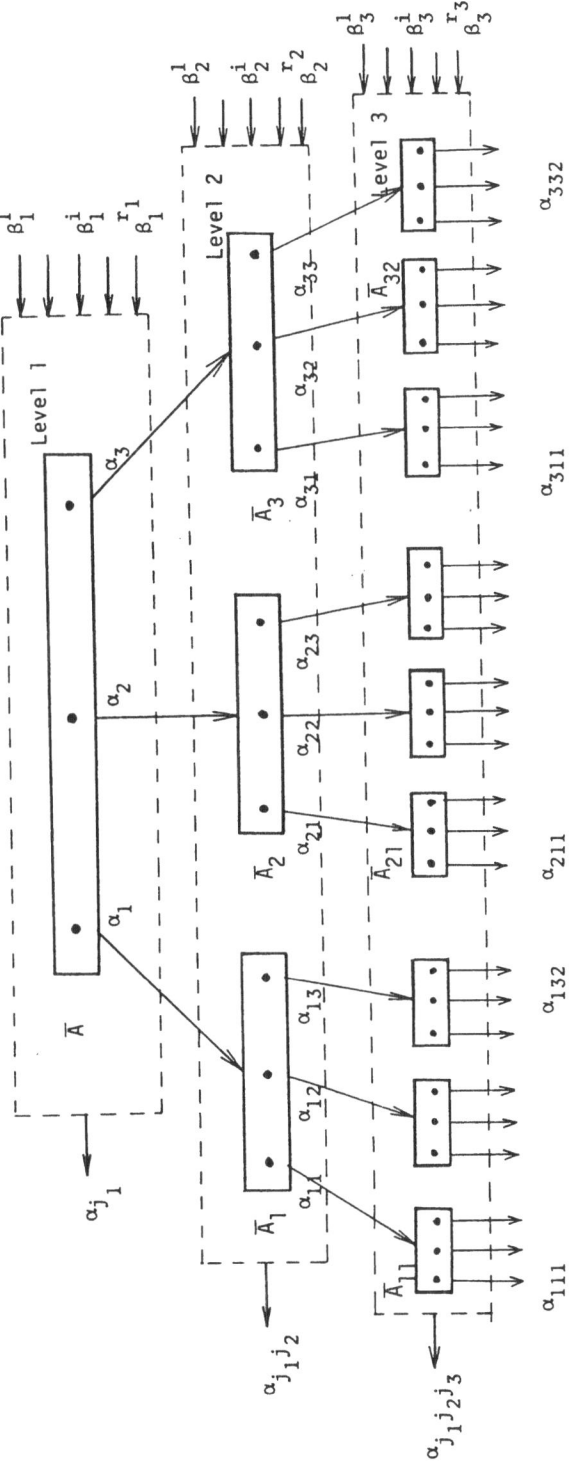

Figure 27   Hierarchical Structure Stochastic Automata Operating in the Multi-Teacher Environments

Ist Level

.Assume that the j_1th action α_{j_1} is selected by the first level automaton $\bar{A}$ at time t and the environmental responses are β_1^i ($i = 1,\ldots,r_1$).

Then,

$$P_{j_1}(t+1) = P_{j_1}(t) + L_1(t)(1 - P_{j_1}(t))$$

$$P_{i_1}(t+1) = P_{i_1}(t)(1 - L_1(t)) \quad (i_1 \neq j_1 ; i_1 = 1,\ldots,r)$$

Nth Level   ($N = 2,\ldots,N$)

Assume that the actions α_{j_1}, $\alpha_{j_1 j_2}$, ..., and $\alpha_{j_1 j_2 \cdots j_{N-1} j_N}$ are selected by the automata $\bar{A},\ldots,\bar{A}_{j_1\cdots j_{N-1}}$ at time t and the environmental responses at the Nth level are β_N^i ($i = 1,\ldots,r_N$).

Then,

$$P_{j_1 j_2 \cdots j_N}(t+1) = P_{j_1 \cdots j_N}(t) + L_N(t)(1 - P_{j_1 \cdots j_N}(t))$$

$$P_{j_1 j_2 \cdots j_{N-1} i_N}(t+1) = P_{j_1 \cdots j_{N-1} i_N}(t)(1 - L_N(t))$$

$$(i_N \neq j_N ; i_N = 1,\ldots,r)$$

All other state probabilities at the Nth level of other automata remain unchanged.

Here, $L_i(t)$ ($i = 1,\ldots,N$) are constructed as follows.

Let

$$\bar{\eta}_s(t) = 1 - \frac{\beta_s^1 + \ldots + \beta_s^{r_s}}{r_s} \quad (s = 1,\ldots,N) .$$

Then, $$L_1(t) = \sum_{s=1}^{N} \bar{\eta}_s(t)\lambda_s ,$$

$$\vdots$$

$$L_n(t) \ = \ \frac{\displaystyle\sum_{s=n}^{N} \bar{\eta}_s(t)\lambda_s}{P_{j_1}(t+1) \ . \ . \ . \ P_{j_{n-1}}(t+1)}$$

We are able to get the following theorem under the condition analogous to that given in [R1]. Since it can be easily proved by the same procedure as done in [R1], we will omit its proof.

Theorem   Assume that there exists a unique path called the optimal path in the hierarchical tree such that at each level, the average sum of the penalty probabilities in the optimal path corresponds to the minimum of those at that level. This means:

Let $\phi_{j_1^* \ldots j_N^*}$ be the optimal path.

Then,
$$E\{ \frac{\beta_s^{1,j_s^*} + \ldots + \beta_s^{r_s,j_s^*}}{r_s} \} \ < \ E\{ \frac{\beta_s^{1,j} + \ldots + \beta_s^{r_s,j}}{r_s} \}$$

for all s ($s = 1,\ldots,N$) and all j ($j = 1,\ldots,r$).

($\beta_s^{i,j}$ ($i=1,\ldots,r_s$; $j=1,\ldots,r$) is the response from the ith teacher at the s level when jth action has been actuated.)

Further, assume that $$0 < \sum_{s=1}^{N} \lambda_s < 1$$

Then,
$$\Delta\pi_{j_1^* \ldots j_N^*}(t) \ = \ E[\ \pi_{j_1^* \ldots j_N^*}(t+1) - \pi_{j_1^* \ldots j_N^*}(t) \ | \ P(t)\ ] \ > \ 0$$

in the open simplex of the action probabilities.

($\pi_{j_1,\ldots,j_N} \ = \ P_{j_1}(t)P_{j_1 j_2}(t) \ldots P_{j_1 j_2 \ldots j_N}(t)$)

The above theorem can be considered as an extension of the theorem 3.1 given in [R1]. Therefore, we could say that the absolute expediency in the general multi-teacher environments can be ensured by the above learning algorithm of the hierarchical structure automata.

EPILOGUE

Throughout this monograph, I have mainly concerned with the learning
behaviors of stochastic automata operating in the multi-teacher environment.
In our daily life, we often encounter the problem in which we need an intelligent behavior in the situation where one action elicits multi-responses from
the unknown multi-criteria environment. Although this problem cannot be easily tackled, active research efforts should be directed to solve it. It is
my hope that this monograph will play an important role to stimulate discussions
in that direction.

This monograph has not been intended to be an encyclopedic treatise in
the area of learning automata. If you are interested in the literature of
learning automata, you should consult the book written by Lakshmivarahan [L6]
(or the survey papers by Narendra et al ([N3], [N4], and [N9])).

REFERENCES

A1 A.O. Allen, <u>Probability, Statistics, and Queueing Theory</u>, Academic Press, 1978.

A2. H. Aso and M. Kimura, "The structures of automata to adapt to an unknown environment", IEEE Trans. Systems, Man, and Cybernetics, Vol. 6, pp. 494-504, 1976.

A3 H. Aso, "A characterization of learning automata", Proceedings of the International Conference on Cybernetics and Society, Tokyo, pp. 1487-1491, 1978.

A4 H. Aso and M. Kimura, "Absolute expediency of learning automata", Information Sciences, Vol. 17, pp. 91-112, 1979.

A5 R.C. Atkinson, G.H. Bower, and E.J. Crothers, <u>An Introduction to Mathematical Learning Theory</u>, Wiley, 1965.

A6 R.J. Aumann and M. Maschler, "Repeated games with incomplete information. The zero-sum extensive case", Report to the U.S. Arms Control and Disarmament Agency, Washington, D.C.: final report on contract ACDA/ST-143 prepared by MATHEMATICA, Princeton, pp. 25-108, 1968.

B1 N. Baba and Y. Sawaragi, "On the learning behavior of stochastic automata under a nonstationary random environment", IEEE Trans. Systems, Man, and Cybernetics, Vol. 5, pp. 273-275, 1975.

B2 N. Baba, "Learning behavior of stochastic automata in the last stage of learning", Information Sciences, Vol. 9, pp. 315-322, 1975.

B3 N. Baba, T. Shoman, and Y. Sawaragi, "A modified convergence theorem for a random optimization theorem", Information Sciences, Vol. 13, pp. 159-166, 1977.

B4   N. Baba, "Theoretical considerations of the parameter self-optimization by stochastic automata", International Journal of Control, Vol. 27, pp. 271-276, 1978.

B5   N. Baba, T. Soeda, T. Shoman, and Y. Sawaragi, "An application of the stochastic automaton to the investment game", International Journal of Systems Science, Vol. 11, pp. 1447-1457, 1980.

B6   N. Baba, "Convergence of a random optimization method for constrained optimization problems", Journal of Optimization Theory and Applications, Vol. 33, pp. 451-461, 1981.

B7   N. Baba, "The absolutely expedient nonlinear reinforcement schemes under the unknown multi-teacher environment", IEEE Trans. Systems, Man, and Cybernetics, Vol. 13, pp. 100-108, 1983.

B8   N. Baba, "On the learning behaviors of variable-structure stochastic automaton in the general n-teacher environment", IEEE Trans. Systems, Man, and Cybernetics, Vol. 13, pp. 224-231, 1983.

B9   N. Baba, "An absolutely expedient nonlinear reinforcement scheme under a nonstationary multi-teacher environment and its applications to practical problems", Proceedings of the Third Yale Workshop on Applications of Adaptive Systems Theory, pp. 110-113, 1983.

B10 N. Baba, "Learning behaviors of stochastic automata and some applications", Working Paper, WP-83-119, IIASA, Austria, pp. 1-26, 1983.

B11 Y. Bar-Shalom and E. Tse, "Caution, probing, and the value of information in the control of uncertain systems", Annals of Economic and Social Measurement , pp. 323-337, 1976.

B12 R. Bellman, Dynamic Programming, Princeton Univ. Press, 1957.

B13 V.E. Benes, Mathematical Theory of Connecting Networks and Telephone Traffic , Academic Press, 1965.

B14 J.O. Berger, <u>Statistical Decision Theory</u>, Springer-Verlag, 1980.

B15 D.P. Bersekas, <u>Dynamic Programming and Stochastic Control</u>, 1976.

C1   B. Chandrasekaran and D.W.C. Shen, "On expediency and convergence in
     variable-structure automata", IEEE Trans. Systems, Man, and Cybernetics,
     Vol. 4, pp. 52-60, 1968.

C2   B. Chandrasekaran and D.W.C. Shen, "Adaptation of stochastic automata in
     nonstationary environments", Proc. NEC, Vol. 23, pp. 33-44, 1967.

C3   B. Chandrasekaran and D.W.C. shen, "Stochastic automata games", IEEE Trans.
     Systems, Science and Cybernetics, Vol. 5, pp. 145-149, 1969.

C4   V.K. Chichinadze, "Random search to determine the extremum of the function
     of several variables", Eng. Cybernetics, Vol. 1, pp. 115-123, 1967.

C5   L.D. Cockrell and K.S. Fu, "On search techniques in switching environment",
     Proceedings of the 9th Symposium Adaptive Processes, Austin, Tex., 1970.

C6   T.M. Cover and M.E. Hellman, "Two armed bandit problem with time-invariant
     finite memory", IEEE Trans. Information Theory, Vol. 14, pp. 185-195, 1970.

D1   C. Derman, <u>Finite State Markovian Decision Processes</u>, Academic Press, 1970.

D2   L.P. Devroye, "On the convergence of statistical search", IEEE Trans.
     Systems, Man, and Cybernetics, Vol. 6, pp. 46-56, 1976.

D3   L.P. Devroye, "Probabilistic search as a search selection procedure", IEEE
     Trans. Systems, Man and Cybernetics, Vol. 6, pp. 315-321, 1976.

D4   L.P. Devroye, "A class of optimal performance directed probabilistic autom-
     ata", IEEE Trans. Systems, Man, and Cybernetics, Vol. 6, pp. 777-783, 1976.

D5   L.P. Devroye, "Progressive global random search of continuous functions",
     Mathematical Programming, Vol. 15, pp. 330-342, 1978.

D6   J. Dieudonne, Foundations of Modern Analysis, Academic Press, 1969.

D7   I.C.W. Dixon and G.P. Szego, Editors, Towards Global Optimization, North-Holland, 1975.

D8   I.C.W. Dixon and G.P. Szego, Editors, Towards Global Optimization 2, North-Holland, 1978

D9   A.V. Dobrovidov and R.L. Stratonovich, "Construction of optimal automata functioning in random media", Automation and Remote Control, Vol. 25, pp. 1289-1296, 1964.

D10   J.L. Doob, Stochastic Processes, Wiley, 1953.

D11   A. Dvoretsky, "On stochastic approximation", in Proc. 3rd Berkeley Symp. on Math. Stat. and Probability, Vol.1, pp. 39-55, 1956.

D12   E.B. Dynkin, Markov Processes, Springer-Verlag, 1965.

E1   Y.M. El-Fattah and C. Foulard, Learning Systems: Decision, Simulation, and Control, Springer-Verlag, 1978.

E2   Y.M. El-Fattah, "Stochastic automata modelling of certain problems of collective behavior", IEEE Trans. Systems, Man, and Cybernetics, Vol. 10, pp. 304-314, 1980.

F1   Yu.A. Flerov, "Some class of multi-input automata", Journal of Cybernetics, Vol. 2, pp. 112-122, 1972.

F2   K.S. Fu and G.J. McMurtry, "A study of stochastic automata as models of adaptive and learning controllers", Purdue Univ., Tech. Rep., TR-EE 65-8, 1965.

F3   K.S. Fu and R.W. Mclaren, "An application of stochastic automata to the synthesis of learning systems", Purdue Univ., Tech. Rep., TR-EE 65-17, 1965.

F4   K.S. Fu and Z.J. Nikolic, "On some reinforcement techniques and their relation to the stochastic approximation", IEEE Trans. Automatic Control, Vol. 11, pp. 756-758, 1966.

F5   K.S. Fu, Sequential Methods in Pattern Recognition and Machine Learning, Academic Press, 1968.

F6   K.S. Fu and T.J. Li, "Formulation of learning automata and automata games", Information Sciences, pp. 237-256, 1969.

F7   K.S. Fu, "Learning control systems - Review and outlook", IEEE Trans. Automatic Control, Vol. 15, pp. 210-221, 1970.

F8   K.S. Fu, Editor, Learning Systems, The American Society of Machanical Engineers, 1973.

F9   K. Fukunaga, Introduction to Statistical Pattern Recognition, Academic Press, 1972.

G1   I.I. Gihman and A.V. Skorohod, Stochastic Differential Equations, Springer-Verlag, 1972.

G2   S.L. Ginsburg, V.Y. Krylov, and M.L. Tsetlin, "On one example of a game of many identical automata", Automation and Remote Control, Vol. 25, pp. 608-612, 1964.

G3   S.L. Ginsburg and M.L. Tsetlin, "Some examples of simulation of the collective behavior of automata", Probl. Peredachi Informatsii, Vol. 1, pp. 54-62, 1965.

G4   A. Ginzburg, Algebraic Theory of Automata, Academic Press, 1968.

G5   E.G. Gladyshev, "On stochastic approximation", Theory of Probability and its Applications, Vol. 10, pp. 275-278, 1965.

G6   R.M. Glorioso, G.R. Grueneich, and J.C. Dunn, "Self organization and
      adaptive routing for communication networks", 1969 EASCON Rec., pp. 243-250.

G7   R.M. Glorioso and G.R. Grueneich, "A training algorithm for systems
      described by stochastic transition matrices", IEEE Trans. Systems, Man,
      and Cybernetics, Vol. 1, pp. 86-87, 1971.

G8   M. Grauer, A. Lewandowski, and A.P. Wierzbicki, Editors, Multiobjective
      and Stochastic Optimization, CP-82-S12, IIASA, AUSTRIA, 1982.

H1   Y.Y. Haimes, W.A. Hall, and H.T. Friedman, Multiobjective Optimization
      in Water Resources Systems, The Surrogate Worth Trade-off Method, Elsivier
      Scientific, 1975.

H2   J.C. Harsanyi, "Games with incomplete information played by "Bayesian"
      players, I", Management Science, Vol. 14, pp. 159-182, 1967.

H3   J.C. Harsanyi, "Games with incomplete information played by "Bayesian"
      players, part II. Bayesian equilibrium points", Ibid, Vol. 14, pp. 320-334,
      1967.

H4   J.C. Harsanyi, "Games with incomplete information played by "Bayesian"
      players, part III. The basic probability distribution of the game",
      Ibid, Vol. 14, pp. 486-502, 1968.

H5   U. Herkenrath, D. Kalin, and S. Lakshmivarahan, "On a general of absorb-
      ing-barrier learning algorithms", Information Sciences, Vol. 24, pp. 255-
      263, 1981.

H6   R.A. Howard, Dynamic Programming and Markov Processes, M.I.T. Press, 1960.

I1   M. Iosifescu and R. Theodorescu, Random Processes and Learning, Springer-
      Verlag, 1969.

I2   K. Ito, Probability Theory, in Japanese, Iwanami, 1952.

I3   S. Ito, An Introduction to Lebesgue Integral, in Japanese, Shokabo, 1963

J1   R.A. Jarvis, "Adaptive global search in a time-invariant environment using a probabilistic automaton", Proc. IREE, Australia, pp. 210-226, 1969.

J2   R.A. Jarvis, "Adaptive global search in a time-variant environment using a probabilistic automaton with pattern recognition supervision", IEEE Trans. Systems, Science and Cybernetics, Vol. 6, pp. 209-217, 1970.

K1   R.L. Kashyap, "Application of stochastic approximation" in Adaptive, Learning and Pattern Recognition Systems, J.M. Mendel and K.S. Fu, Editors, Academic Press, New York, 1970.

K2   R.L. Kashyap, "Syntactic decision rules for recognition of spoken words and phrases using a stochastic automaton", IEEE Trans. Pat. Analys. and Mach. Intel., Vol. PAMI-1, No. 2, pp.154-163, 1979.

K3   V.Y. Katkovnik and I.V. Antonov, "Generalization of the concept of statistical gradient", Automation and Remote Control, No. 6, pp. 26-33, 1972.

K4   J.G. Kemeny and J.L. Snell, Finite Markov Chains, Springer-Verlag, 1976.

K5   D.E. Koditschek and K.S. Narendra, "Fixed structure automata in multi-teacher environment", IEEE Trans. Systems, Man, and Cybernetics, Vol. 7, pp. 616-624, 1977.

K6   E. Kohlberg, "On the nucleolus of a characteristic game", SIAM J. Appl. Math., Vol. 20, pp. 62-66, 1971.

K7   E. Kohlberg, "Optimal strategies in repeated games with incomplete information", International Journal of Game Theory, Vol. 4, pp. 7-24, 1974.

K8   V.I. Krinskii, "An asymptotically optimal automaton with exponential convergence", Bio Physics, Vol. 9, pp. 484-487, 1964.

K9   V.Y. Krylov, "On one stochastic automaton which is asymptotically optimal in a random media", Automation and Remote Control, Vol. 24, pp. 1114-1116, 1963.

K10   V.Y. Krylov and M.L. Tsetlin, "Games between automata", Automation and Remote Control, Vol. 24, pp. 889-900, 1963.

K11   H.J. Kushner, Stochastic Stability and Control, Academic Press, 1967.

K12   H.J. Kushner, Introduction to Stochastic Control, Holt, Rinehart and Winston, 1971.

K13   H.J. Kushner, M.A.L. Thathachar, and S. Lakshmivarahan, "Two-state automaton with linear reward-inaction reinforcement scheme - A counter example", IEEE Trans. Systems, Man and Cybernetics, Vol. 2, pp. 292-294, 1972.

K14   H.J. Kushner, "Stochastic approximation algorithms for the local optimization of functions with nonunique stationary points", IEEE Trans. Automatic Control, Vol. 17, pp. 646-654, 1972.

K15   H.J. Kushner, "Convergence of recursive adaptive and identification procedures via weak convergence theory", IEEE Trans. Automatic Control, Vol. 22, pp. 921-930, 1977.

L1   S. Lakshmivarahan and M.A.L. Thathachar, "Absolutely expedient learning algorithms for stochastic automata", IEEE Trans. Systems, Man, and Cybernetics, Vol. 3, pp. 281-286, 1973.

L2   S. Lakshmivarahan, "Learning algorithms for stochastic automata acting in nonstationary random environments", Journal of Cybernetics, Vol. 4, pp. 73-85, 1974.

L3   S. Lakshmivarahan and and M.A.L. Thathachar, "Absolute expediency of Q-
      and S- model learning algorithms", IEEE Trans. Systems, Man and Cybernet-
      ics, Vol. 6, pp. 222-226, 1976.

L4   S. Lakshmivarahan and M.A.L. Thathachar, "Bounds on the probability of
      convergence of learning automata", IEEE Trans. Systems, Man and Cybernet-
      ics, Vol. 6, pp. 756-763, 1976.

L5   S. Lakshmivarahan and K.S. Narendra, "Learning algorithms for two person
      zero sum stochastic games with incomplete information", Mathematics of
      Operations Research, Vol. 6, 1981.

L6   S. Lakshmivarahan, Learning Algorithms Theory and Applications, Springer-
      Verlag, 1981.

L7   L. Ljung, "Analysis of recursive stochastic algorithms", IEEE Trans.
      Automatic Control, Vol. 22, pp. 551-575, 1977.

L8   M. Loeve, Probability Theory, 4th Edition, Springer-Verlag, 1977.

L9   R.D. Luce and H. Raiffa, Games and Decisions, Wiley, 1957.

M1   P. Mars and M.S. Chrystall, "Real-time telephone traffic simulation using
      learning automata routing", S & IS Report No. 7909, Dept. of Eng. and
      Applied Science, Yale University, 1979.

M2   P. Mars, K.S. Narendra, and M. Crystall, "Learning automata control of
      computer communication networks", Proceedings of the Third Yale Workshop
      on Applications of Adaptive Systems Theory, pp. 114-119, 1983.

M3   L.G. Mason, "Self-optimizing allocation systems", Ph.D. University of
      Saskatchewan, Canada, 1972.

M4   L.G. Mason, "An optimal learning algorithm for S-model environments", IEEE Trans. Automatic Control, Vol. 18, pp. 493-496, 1973.

M5   R.W. McLaren, "A stochastic automaton model for synthesis of learning systems", IEEE Trans. Systems, Science and Cybernetics, Vol. 2, pp. 109-114, 1966.

M6   G.J. McMurtry and K.S. Fu, "A variable-structure automaton used as a multimodal search technique", IEEE Trans. Automatic Control, Vol. 11, pp. 379-387, 1966.

M7   N. Megiddo, "On repeated games with incomplete information played by non-bayesian players", International Journal of Game Theory, Vol. 9, pp. 157-167, 1979.

M8   J.M. Mendel and K.S. Fu, Editors, Adaptive, Learning and Pattern Recognition Systems, Academic Press, 1970

M9   J.M. Mendel, Discrete Techniques of Parameter Estimation, Dekker, 1973.

M10   J.M. Mendel, "Reinforcement learning models and their applications to control problems: Learning Systems", 1973 Joint Automatic Control Conference Proceedings.

M11   J.F. Mertens and S. Zamir, "The value of two-person zero-sum repeated games with lack of information on both sides", International Journal of Game Theory, Vol. 1, pp. 39-64, 1971.

M12   J.F. Mertens and S. Zamir, "Minmax and maxmin of repeated games with incomplete information", International Journal of Game Theory, Vol. 9, pp. 201-215, 1979.

M13   M.R. Meybodi and S. Lakshmivarahan, "A learning approach to priority assignment in a two class M/M/1 queueing system with unknown parameters", Proceedings of the Third Yale Workshop on Applications of Adaptive Systems Theory, pp. 106-109, 1983.

N1   K.S. Narendra and R. Viswanathan, "A two-level system of stochastic autom-
     ata for periodic random environments", IEEE Trans. Systems, Man, and
     Cybernetics, Vol. 2, pp. 285-289, 1972.

N2   K.S. Narendra and R. Viswanathan, "Learning models using stochastic autom-
     ata", in Proc. 1972 Int. Conf. Cybernetics and Society, Washington, D.C.,
     pp. 9-12, 1972.

N3   K.S. Narendra and M.A.L. Thathachar, "Learning automata - a survey", IEEE
     Trans. Systems, Man, and Cybernetics, Vol. 4, pp. 323-334, 1974.

N4   K.S. Narendra and S. Lakshmivarahan, "Learning automata - a critique",
     Tech. Rep. 7703, Yale University, 1977.

N5   K.S. Narendra, E. Wright, and L.G. Mason, "Application of learning autom-
     ata to telephone traffic routing", IEEE Trans. Systems, Man, and Cybenet-
     ics, pp. 785-792, 1977.

N6   K.S. Narendra and M.A.L. Thathachar, "On the behavior of learning autom-
     aton in a changing environment with routing applications", IEEE Trans.
     Systems, Man, and Cybernetics, Vol. 10, pp. 262-269, 1980.

N7   K.S. Narendra, "The use of learning algorithms in telephone traffic rout-
     ing - a methodology", Tech. Rep. 8203, Yale University, 1982.

N8   K.S. Narendra and R.M. Wheeler, "An n-player sequential stochastic game
     with identical payoffs", Tech. Rep. 8209, Yale University, 1982.

N9   K.S. Narendra, "Recent developments in learning automata - theory and
     applications", Proceedings of the Third Yale Workshop on Applications of
     Adaptive Systems Theory, pp. 90-99, 1983.

N10 O.V. Nedzelnitsky, "Learning automata routing in data communication net-
     works", Proceedings of the Third Yale Workshop on Applications of Adapt-
     ive Systems Theory, pp. 142-147, 1983.

N11   J. Neumann and O. Morgenstern, Theory of Games and Economic Behavior, Princeton University Press, 1953.

N12   M.B. Nevelson and R.Z. Has'minskii, Stochastic Approximation and Recursive Estimation, Translation of the American Mathematical Society, 1973.

N13   N.J. Nilsson, Learning Machines, McGraw-Hill, 1965.

N14   M.F. Norman, "Some convergence theorems for stochastic learning models with distance diminishing operators", Journal of Mathematical Psychology, Vol. 5, pp. 61-101, 1968.

N15   M.F. Norman, "On linear models with two absorbing barriers", Journal of Mathematical Psychology, Vol. 5, pp. 225-241, 1968.

N16   M.F. Norman, "Slow learning", The British Journal of Mathematical and Statistical Psychology, Vol. 21, pp. 141-159, 1968.

N17   M.F. Norman, Markov Processes and Learning Models, Academic Press, 1972.

N18   M.F. Norman, "A central limit theorem for Markov processes that move by small steps", The Annals of Probability, Vol. 2, pp. 1065-1074, 1974.

N19   M.F. Norman, "Markovian learning process", SIAM Review, Vol. 16, pp. 143-162, 1974.

N20   M.F. Norman, "Approximation of stochastic processes by Gaussian diffusions and applications to Wright-Fisher generic model.", SIAM Journal of Applied Mathematics, Vol. 29, pp. 225-242, 1975.

P1   K.R. Parthasarathy, Probability Measures in Metric Spaces, Academic Press, 1965.

P2   B.T. Poljak, "Nonlinear programming methods in the presence of noise", Mathematical Programming, pp. 87-97, 1978.

P3   V.A. Ponomarev, "A construction of an automaton which is asymptotically optimal in a stationary random media", Bio Physics, Vol. 9, pp.104-110, 1964.

P4   J.P. Ponssard and S. Zamir, "Zero-sum sequential games with incomplete information", International Journal of Game Theory, Vol. 2, pp. 99-110, 1974.

P5   J.P. Ponssard, "Zero-sum games with "almost" perfect information", Management Science, Vol. 21, pp. 794-805, 1975.

P6   A.S. Poznyak, "Investigation of convergence of algorithms for learning stochastic automata", Automation and Remote Control, pp. 77-91, 1973.

P7   A.S. Poznyak, "Learning automata in stochastic programming problem", Automation and Remote Control, pp. 1608-1619, 1973.

R1   K.R. Ramakrishnan, "Hierarchical Systems and Co-operative Games of Learning Automata", Ph.D. Thesis, Indian Institute of Science, Bangalore, India, 1982.

R2   J.S. Riordon, "Optimal feedback characteristics from stochastic automaton models", IEEE Trans. Automatic Control, Vol. 14, pp. 89-92, 1969.

R3   J.S. Riordon, "An adaptive automaton controller for discrete-time Markov processes", Automatica, Vol. 5, pp. 721-730, 1969.

R4   H. Robbins and S. Monro, "A stochastic approximation method", Annals of Mathematical Statistics, Vol. 22, pp. 400-407, 1951.

R5   H. Robbins, "Sequential decision problem with finite memory", Proceedings of the National Academy of Sciences, Vol. 42, pp. 920-923, 1956.

R6   H.L. Royden, Real Analysis, Macmillan Pub. Co. Inc., 1963.

S1   S.M. Samuels, "Randomized rules for the two-armed bandit with finite memory", Annals of Mathematical Statistics, Vol. 39, pp. 2103-2107, 1968.

S4   G.N. Saridis, Self-Organizing Control of Stochastic Systems, Marcel Dekker Inc., 1978.

S3   Y. Sawaragi and N. Baba, "A note on the learning behavior of variable-structure stochastic automata", IEEE Trans. Systems, Man, and Cybernetics, Vol. 3, pp. 644-647, 1973.

S4   Y. Sawaragi and N. Baba, "Two ε-optimal nonlinear reinforcement schemes for stochastic automata", IEEE Trans. Systems, Man, and Cybernetics, Vol. 4, pp. 126-131, 1974.

S5   Y. Sawaragi, N. Baba, and T. Soeda, "New topics of learning automata", Journal of Cybernetics and Information Science, Vol. 1, pp. 112-120, 1977.

S6   I.J. Shapiro and K.S. Narendra, "Use of stochastic automata for parameter self-optimization with multimodal performance criteria", IEEE Trans. Systems Science and Cybernetics, Vol. 5, pp. 352-360, 1969.

S7   J. Sklansky, "Learning systems for automatic control", IEEE Trans. Automatic Control, Vol. 11, pp. 6-19, 1966.

S8   C.V. Smith and R. Pyke, "The Robbins-Isbell two armed bandit problem with finite memory", Annals of Mathematical Statistics, Vol. 36, pp. 1375-1386, 1965.

S9   V.G. Sragovich, "Automata with multivalued input and their behavior in random environments", Journal of Cybernetics, Vol. 2, pp. 79-108, 1972.

S10 P.R. Srikantakumar and K.S. Narendra, "A learning model for routing in telephone networks", SIAM Journal on Control and Optimization, Vol. 20, pp. 34-57, 1982.

S11 P.R. Srikantakumar, "Application of learning theory to communication networks control", Proceedings of the Third Yale Workshop on Applications of Adaptive Systems Theory, pp. 135-141, 1983.

T1   M.A.L. Thathachar and R. Bhakthavathsalam, "Learning automaton operating in parallel environments", Journal of Cybernetics and Information Science, Vol. 1, pp. 121-127, 1978.

T2   M.A.L. Thathachar and K.R. Ramakrishnan, "An automaton model of a hierarchical learning system", IFAC Congress, Kyoto, Japan, 1981.

T3   M.A.L. Thathachar and P.S. Sastry, "A new approach to the design of reinforcement schemes for learning automata", Tech. Rep., Indian Institute of Science, 1983.

T4   M.L. Tsetlin, "On behavior of finite automata in random media", Automation and Remote Control, Vol. 22, pp. 1345-1354, 1961.

T5   M.L. Tsetlin, Automaton Theory and Modelling of Biological Systems, Academic Press, 1963.

T6   H. Tsuji, M. Mizumoto, J. Toyoda, and K. Tanaka, "An automaton in the nonstationary random environment", Information Sciences, Vol. 6, pp. 123-142, 1973.

T7   Y.Z. Tsypkin, Adaptation and Learning in Automatic Systems, Academic Press, 1971.

T8   Y.Z. Tsypkin and A.S. Poznyak, "Finite learning automata", Engineering Cybernetics, Vol. 10, pp. 478-490, 1972.

T9   Y.Z. Tsypkin, Foundations of the Theory of Learning Systems, Academic Press, 1973.

V1   V.I. Varshavskii and I.P. Vorontsova, "On the behavior of stochastic automata with variable structure", Automation and Remote Control, Vol. 24, pp. 327-333, 1963.

V1   E.M. Vaisbord, "Game of two automata with differing memory depths",
      Automation and Remote Control, Vol. 29, pp. 440-451, 1968.

V2   E.M. Vaisbord, "Game of many automata with various depths of memory",
      Automation and Remote Control, Vol. 29, pp. 1938-1943, 1968.

V3   V.I. Varshavskii and I.P. Vorontsova, "On the behavior of stochastic auto-
      mata with variable structure", Automation and Remote Control, Vol. 24,
      pp. 327-333, 1963.

V4   V.I. Varshavskii, M.V. Meleshina, and M.L. Tsetlin, "Behavior of automata
      in periodic random media and the problem of synchronization in the presence
      of noise", Probl. Peredachi Informatsii, Vol. 1, pp. 65-71, 1965.

V5   V.I. Varshavskii, "Some effects in the collective behavior of automata",
      Machine Intelligence, B. Meltzer and D. Michie, Editors, Edinburgh:
      Edinburgh Univ., 1969.

V6   V.I. Varshavskii, "Automata games and control problems", IFAC Congress,
      Paris, 1972.

V7   R. Viswanathan and K.S. Narendra, "Application of stochastic automata
      models to learning systems with multimodal performance criteria", Tech.
      Rep. CT 40, Yale University, 1971.

V8   R. Viswanathan and K.S. Narendra, "A note on linear reinforcement scheme
      for variable structure stochastic automata", IEEE Trans. Systems, Man,
      and Cybernetics, Vol. 2, pp. 292-294, 1972.

V9   R. Viswanathan and K.S. Narendra, "Competitive and cooperative games of
      variable-structure stochastic automata", Joint Automatic Control Conf.,
      1972.

V10 R. Viswanathan and K.S. Narendra, "Stochastic automata models with application to learning systems", IEEE Trans. Systems, Man, and Cybernetics, Vol. 3, pp. 107-111, 1973.

V11 R. Viswanathan and K.S. Narendra, "Games of stochastic automata", IEEE Trans. Systems, Man, and Cybernetics, Vol. 4, pp. 131-135, 1974.

V12 I.P. Vorontsova "Algorithms for changing automaton transition probabilities", Problemi Peredachi Informatsii, Vol. 1, pp. 122-126, 1965.

W1 M.D. Waltz and K.S. Fu, "A heuristic approach to reinforcement learning control system", IEEE Trans. Automatic Control, Vol. 10, pp. 390-398, 1965.

W2 M.T. Wasan, Stochastic Approximations, Cambridge University Press, 1969.

W3 S. Watanabe, Knowing and Guessing, Wiley, 1969.

W4 S. Watanabe, "Creative learning and propensity automaton", IEEE Trans. Systems, Man, and Cybernetics, Vol. 5, pp. 603-610, 1975.

W5 W.G. Wee and K.S. Fu, "A heuristic approach to reinforcement learning control system", IEEE Trans. Systems, Man, and Cybernetics, Vol. 5, pp. 215-223, 1969.

W6 I.H. Witten, "Finite time performance of some two-armed bandit controller", IEEE Trans. Systems, Man, and Cybernetics, Vol. 3, pp. 194-197, 1973.

W7 I.H. Witten, "The apparent conflict between estimation and control - A survey of two-armed bandit problem", Journal of Franklin Institute, Vol. 301, pp. 161-189.

W8 I.H. Witten, "An adaptive optimal controller for discrete-time Markov environments", Information and Control, Vol. 34, pp. 286-295, 1977.

Z1 L.A. Zadeh, "Fuzzy sets", Information and Control, Vol. 8, pp. 338-353, 1965.

Z2 S. Zamir, "On the notion of the value for games with infinitely many stages", Annals of Statistics, Vol. 1, pp. 791-796, 1973.

INDEX

absolutely expedient	2, 3, 8
absolutely expedient algorithm	10
absolute expediency in the general n-teacher environment	3 ,23 ,27
a priori information	1
asymptotically optimal	2
average penalty	6
average sum of the penalty probabilities	107
average weighted reward in the general n-teacher environment	22, 29
Basic space	12, 56
Borel field	12, 56
coalition	4, 91, 94, 96
conditional expectation	13, 29, 60
converges with probabiliti 1	1, 16, 34
cooperative game	90, 94
discrete parameter Semi-Martingale	14
distribution function	13, 60, 74
expedient	7
expedient in the general n-teacher environment	23
fixed-structure automata	18, 21, 36

GAE reinforcement scheme	3, 25, 26, 27, 37
Gaussian white noise	72, 78
general n-teacher environment	3, 18, 21, 55
GL_{R-I} scheme	26, 37, 64, 65
GNA scheme	26, 37
hierarchical structure automata	4, 90, 93, 104
incomplete information	90
learning automaton	1, 56
learning behavior	1, 17, 55, 71, 104
learning control	1
L_{R-I} scheme	9
majority decision	36
mathematical expectation	13, 16, 32
MGAE scheme	4, 55, 59, 76, 78
multi-teacher environment	1, 17, 25, 55, 71, 104
noise-corrupted, multi-objective function	4, 71, 78
nonstationary multi-teacher environment	3, 55, 56, 71, 88
nonstationary random environment	6
optimal control	1
optimality	7, 8
ε-optimality	2, 3, 7

optimal in the general n-teacher environment   24

ε-optimal in the general n-teacher environment   3, 24, 27

optimal in NMT   58

ε-optimal in NMT   58

parameter self-optimization   4, 71

Pareto-optimal parameter   77

payoff   91

penalty response   4, 9, 18, 25, 74

penalty strength   58

perfect information   1

P-model   4, 20, 55, 63, 69, 73, 104

probability density function   13

probability measure space   12, 15, 56, 74

Q-model   5, 20

random environment   1, 5, 6

random variable   12

reinforcement scheme   3, 6

repeated game   90, 91

reward-inaction scheme   9

reward response   4, 18, 25

routing of messages in communication network   17

Semi-Martingale   14, 16, 32

single teacher environment   4, 6, 23

smallest Borel field   15, 56, 74

S-model   5, 20, 56, 64

stationary random environment   1, 6

stochastic automaton   1, 4, 17, 55, 72, 91, 104

stochastic process   14

sum of the penalty strength   58

super regular function   34

telephone network routing   70

two person zero-sum game   71

uniformly distributed random variable   64

variable-structure stochastic automaton   6

Lecture Notes in Control and Information Sciences

Edited by M. Thoma

Vol. 43: Stochastic Differential Systems
Proceedings of the 2nd Bad Honnef Conference
of the SFB 72 of the DFG at the University of Bonn
June 28 – July 2, 1982
Edited by M. Kohlmann and N. Christopeit
XII, 377 pages. 1982.

Vol. 44: Analysis and Optimization of Systems
Proceedings of the Fifth International
Conference on Analysis and Optimization of Systems
Versailles. December 14–17, 1982
Edited by A. Bensoussan and J. L. Lions
XV, 987 pages, 1982

Vol. 45: M. Arató
Linear Stochastic Systems
with Constant Coefficients
A Statistical Approach
IX, 309 pages. 1982

Vol. 46: Time-Scale Modeling of Dynamic Networks
with Applications to Power Systems
Edited by J. H. Chow
X, 218 pages. 1982

Vol. 47: P. A. Ioannou, P. V. Kokotovic
Adaptive Systems with Reduced Models
V, 162 pages. 1983

Vol. 48: Yaakov Yavin
Feedback Strategies for Partially
Observable Stochastic Systems
VI, 233 pages, 1983

Vol. 49: Theory and Application of Random Fields
Proceedings of the IFIP-WG 7/1
Working Conference
held under the joint auspices of the
Indian Statistical Institute
Bangalore, India, January 1982
Edited by G. Kallianpur
VI. 290 pages. 1983

Vol. 50: M. Papageorgiou
Applications of Automatic Control Concepts
to Traffic Flow Modeling and Control
IX, 186 pages. 1983

Vol. 51: Z..Nahorski, H.F. Ravn, R.V.V. Vidal
Optimization of Discrete Time Systems
The Upper Boundary Approach
V, 137 pages 1983

Vol. 52: A. L. Dontchev
Perturbations, Approximations and Sensitivity Analysis
of Optimal Control Systems
IV, 158 pages. 1983

Vol. 53: Liu Chen Hui
General Decoupling Theory of Multivariable
Process Control Systems
XI, 474 pages. 1983

Vol. 54: Control Theory for Distributed
Parameter Systems and Applications
Edited by F. Kappel, K. Kunisch,
W. Schappacher
VII, 245 pages. 1983.

Vol. 55: Ganti Prasada Rao
Piecewise Constant Orthogonal Functions
and Their Application to Systems and Control
VII, 254 pages. 1983.

Vol. 56: Dines Chandra Saha, Ganti Prasada Rao
Identification of Continuous
Dynamical Systems
The Poisson Moment Functional
(PMF) Approach
IX, 158 pages. 1983.

Vol. 57: T. Söderström, P. G. Stoica
Instrumental Variable Methods
for System Identification
VII, 243 pages. 1983.

Vol. 58: Mathematical Theory of
Networks and Systems
Proceedings of the MTNS-83 International
Symposium
Beer Sheva, Israel, June 20–24, 1983
Edited by P. A. Fuhrmann
X, 906 pages. 1984

Vol. 59: System Modelling and Optimization
Proceedings of the 11th IFIP Conference
Copenhagen, Denmark, July 25-29, 1983
Edited by P. Thoft-Christensen
IX, 892 pages. 1984

Vol. 60: Modelling and Performance
Evaluation Methodology
Proceedings of the International Seminar
Paris, France, January 24–26, 1983
Edited by F. Bacelli and G. Fayolle
VII, 655 pages. 1984

Vol. 61: Filtering and Control of Random
Processes
Proceedings of the E.N.S.T.-C.N.E.T. Colloquium
Paris, France, February 23–24, 1983
Edited by H. Korezlioglu, G. Mazziotto, and
J. Szpirglas
V, 325 pages. 1984

Lecture Notes in Control and Information Sciences

Edited by M. Thoma

Vol. 62: Analysis and Optimization ·
of Systems
Proceedings of the Sixth International
Conference on Analysis and Optimization
of Systems
Nice, June 19–22, 1984
Edited by A. Bensoussan, J. L. Lions
XIX, 591 pages. 1984.

Vol. 63: Analysis and Optimization
of Systems
Proceedings of the Sixth International
Conference on Analysis and Optimization
of Systems
Nice, June 19–22, 1984
Edited by A. Bensoussan, J. L. Lions
XIX, 700 pages. 1984.

Vol. 64: Arunabha Bagchi
Stackelberg Differential Games
in Economic Models
VIII, 203 pages, 1984

Vol. 65: Yaakov Yavin
Numerical Studies
in Nonlinear Filtering
VIII, 273 pages, 1985.

Vol. 66: Systems and Optimization
Proceedings of the Twente Workshop
Enschede, The Netherlands, April 16–18, 1984
Edited by A. Bagchi, H. Th. Jongen
X, 206 pages, 1985.

Vol. 67: Real Time Control of Large Scale Systems
Proceedings of the First European Workshop
University of Patras, Greece, Juli 9–12, 1984
Edited by G. Schmidt, M. Singh, A. Titli,
S. Tzafestas
XI, 650 pages, 1985.

Vol. 68: T. Kaczorek
Two-Dimensional Linear Systems
IX, 397 pages, 1985.

Vol. 69: Stochastic Differential Systems –
Filtering and Control
Proceedings of the IFIP-WG 7/1 Working Conference
Marseille-Luminy, France, March 12-17, 1984
Edited by M. Metivier, E. Pardoux
X, 310 pages, 1985.

Vol. 70: Uncertainty and Control
Proceedings of a DFVLR International Colloquium
Bonn, Germany, March, 1985
Edited by J. Ackermann
IV, 236 pages, 1985.

Vol. 71: N. Baba
New Topics in Learning Automata
Theory and Applications
VII, 131 pages, 1985.